Farm Fresh

NORTH CAROLINA

Farm Fresh

NORTH CAROLINA

THE GO-TO GUIDE TO GREAT Farmers' Markets · Farm Stands
Farms · Apple Orchards · U-Picks · Kids' Activities · Lodging · Dining
Choose-and-Cut Christmas Trees · Vineyards and Wineries · and More

Diane Daniel

*To Alice and David,
My dear farm-loving friends,
keep it fresh in NC!
Diane
3/11*

The University of North Carolina Press   Chapel Hill

A **SOUTHERN GATEWAYS** GUIDE
The publication of this book was supported by a grant from the Golden LEAF Foundation.

Library of Congress Cataloging-in-Publication Data
Daniel, Diane.
Farm fresh North Carolina : the go-to guide to great farmers' markets, farm stands, farms, apple orchards, U-picks, kids' activities, lodging, dining, choose-and-cut Christmas trees, vineyards and wineries, and more / Diane Daniel.
p. cm. — (A Southern gateways guide)
"The publication of this book was supported by a grant from the Golden LEAF Foundation."
Includes bibliographical references and index.
ISBN 978-0-8078-7182-9 (pbk : alk. paper)
1. North Carolina — Guidebooks. 2. North Carolina — Tours. 3. Farms — North Carolina — Guidebooks. 4. Farmers' markets — North Carolina — Guidebooks. 5. Agriculture — North Carolina — Guidebooks. I. Title.
F252.3.D36 2011
975.6′044 — dc22 2010032655

15 14 13 12 11 5 4 3 2 1

MIX
Paper from
responsible sources
FSC® C068106

FOR MY PARENTS, KITTY AND DAN DANIEL,
who raised me right, and right here, in North Carolina

AND FOR LINA WESSEL KOK,
my partner in life and all its adventures, near and far

CONTENTS

Ashe
Alleghany
Mitchell
Watauga
Wilkes
Avery
Madison
Yancey
Buncombe
McDowell
Haywood
Swain
Rutherford
Graham
Jackson
Henderson
Cherokee
Macon
Transylvania
Polk
Clay

Surry
Stokes
Rockingham
Yadkin
Forsyth
Guilford
Davie
Davidson
Randolph

MOUNTAINS

Caldwell
Alexander
Iredell
Burke
Catawba
Rowan
Lincoln
Cabarrus
Cleveland
Gaston
Stanly
Mecklenburg
Union
Anson

CHARLOTTE
AREA

TRIANGLE

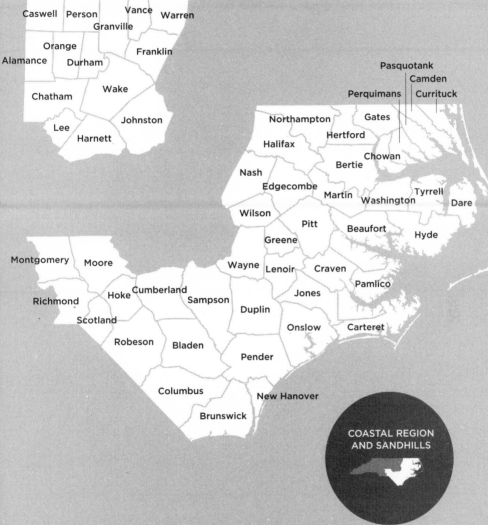

Caswell   Person   Vance   Warren
              Granville
        Orange   Franklin
Alamance   Durham
                    Wake
Chatham
          Lee   Johnston
              Harnett

Pasquotank
                                    Camden
Perquimans   Currituck
              Gates
Northampton
              Hertford
        Halifax   Chowan
                    Bertie
  Nash
        Edgecombe   Martin
                              Washington   Tyrrell
  Wilson   Dare
              Pitt   Beaufort
        Greene   Hyde

Montgomery   Moore
                                    Wayne   Lenoir   Craven
Richmond   Hoke Cumberland   Pamlico
                          Sampson   Jones
Scotland   Duplin
        Robeson   Bladen   Onslow   Carteret
                          Pender

        Columbus   New Hanover
              Brunswick

COASTAL REGION
AND SANDHILLS

INTRODUCTION

At Fickle Creek Farm in Orange County I witnessed a chicken lay an egg. She was hovering a few inches above her nesting box and out it dropped. I'd gathered eggs before, many still warm, but I'd never seen that. Most people haven't, and that even includes some farmers I asked. I couldn't stop talking about it for weeks.

That was just one of the many thrills I experienced while visiting farms across North Carolina. During the course of my research, I kissed a llama, fed an alpaca, patted a giant hog, picked all types of berries, took a hayride out to a pumpkin patch, sipped on merlot while overlooking the grape-vines, shopped at dozens of farmers' markets, watched sorghum syrup being made, stayed on a Christmas tree farm, ate peach ice cream from a picnic table overlooking the orchard, watched goat cheese being made while the goats grazed outside, and enjoyed scrumptious meals sourced from local farms.

Driving to the farms was usually part of the fun. Meeting the farmers often made the experience transformational. They do back-breaking work all day and still find the energy and passion to entertain, educate, and en-lighten us.

I invite you to join me in exploring North Carolina through its family farms, produce stands, farmers' markets, wineries, orchards, and more. I'll show you where to cut a fresh Christmas tree or pick a peck of apples. Want to stay overnight on a working farm and eat a meal with freshly har-vested ingredients? I know just the place.

Farms mean different things to different people. To parents, they might be about showing their children where food comes from. To local-food proponents, farms are the source of their meals. To local-economy advocates, they provide a way of keeping business in the community. To couples, farms offer the perfect outing, such as a visit to one or more of our dozens of wineries.

My love of farms comes from my love of the land. Farmland and farms were part of my landscape when I was growing up outside of Raleigh. My parents, lifelong southerners, moved to the state in 1958, when I was a year old.

They both loved to grow things, and we had a large vegetable garden. Behind it were miles of woods and pastures — my playground. In the summer, we would pick buckets of blackberries. Every December we'd tromp through the woods to find the perfect eastern red cedar, drag it to the house, and decorate it. No offense to the tree farmers in western North Carolina, but I still prefer the cedar over the Fraser fir.

On many occasions, when I wasn't in school, I would join my mom, a nurse who worked for United Cerebral Palsy, as she visited clients at their homes. We drove deep into the country, on paved and dirt roads, passing tobacco farm after tobacco farm, slowing for tractors, and occasionally stopping at a corner market for Nekot cookies and Dr Peppers.

My dad had an office job at Nationwide Insurance, but he lived on a farm in Granville County until he reached adulthood. I remember once watching him compete in a watermelon-seed spitting contest at the State Farmers' Market, back when it was near downtown. I think he came in second.

When I returned to North Carolina in 2003, after an almost thirty-year absence, I discovered, not surprisingly, that things had changed. The rural landscape of my youth had become urbanized, as farmland was being rapidly lost to development.

But I also was delighted to discover that the red clay soil and farmland of my youth were still there, if I looked. And thanks to the tobacco buyout, there were actually more small farms doing interesting things. The end of the federal tobacco support program in 2005 didn't kill the tobacco industry, but it greatly consolidated it. Tobacco farming was no longer lucrative for most family farms. So farmers got creative. They're good at that.

Farm Fresh North Carolina is a celebration of our farmers' ingenuity and successes. The book focuses on goods and services produced directly for the consumer. These might include produce and livestock, Christmas trees, wine, pick-your-own fruit, and even meals and lodging.

Tractors still share the road with
cars in some regions of North Carolina.
Photo by Selina Kok.

This guidebook also introduces a new crop of farmers—young and not-so-young people who want to return to the land, farm sustainably, and support their local economy. While the number of farms statewide has decreased, the number of small, sustainable farms is slowly rising. The renewed farm movement is the bridge between the old-time North Carolina I grew up in and the more progressive state I live in now.

In the past several years, both new and long-standing farms have been boosted by the statewide and national boom in farmers' markets and Community Supported Agriculture (csa) programs, where customers buy a share in the farm's seasonal harvest and receive a box of the farm's produce weekly. The Slow Food and eat-local movements have contributed greatly as well, with many home cooks and restaurant chefs going out of their way to use ingredients from area farmers. Some of those chefs, cooks, and farmers have shared their favorite farm-sourced recipes with us. You'll find a handful in every chapter.

While this book features many farms that are set up to serve the public, others are private, but the farmers nonetheless feel it's important to let people see the work they do.

As organics pioneer Bill Dow at Ayrshire Farm in Chatham County told me, "If people don't learn about where their food comes from, then we're in serious trouble. I feel like it's my duty to show them."

Whatever your reason for visiting North Carolina farms—shopping for food, kissing a llama, tasting wine, or waiting for an egg to drop from a chicken—get out there and enjoy yourself. And thank a farmer while you're at it.

HOW TO USE THIS BOOK

This guidebook is organized around the five regions of the state. The mountain and coastal/Sandhills regions cover the largest territory. The remaining three are clustered around urban areas—Charlotte, the Triad, and the Triangle.

Each region is then divided into categories of interest, which are explained below. If a farm could fit into multiple categories, I chose the one that best represents it. Inside the categories, the listings are organized alphabetically by county. Each chapter includes a map of the region showing the counties and key cities. Please note that because many of these listings are in rural locations, the town the address is in isn't always the closest town to the farm.

All farms and other businesses in the book have some interaction with the public, but this interaction varies greatly from place to place, especially among farms. Some have regular hours while others are open seasonally or occasionally. Some are open only by appointment. Farm schedules are inherently seasonal, so always check ahead before visiting. Most important, please do not drop in on a farm outside of its stated hours. A tip sheet for visiting farms can be found at the end of this introduction.

Farms

This is the broadest of categories. It includes working farms, historic farms, and educational farms. Also under this category are some gardens and nurseries. I chose to include only nurseries that specialize in edible plants or plants native to North Carolina. This book contains both sustainable and organic farms as well as "conventional" farms, which rely more heavily on pesticides. All are family farms of a few to a few hundred acres. What you won't find in this book (I hope) are industrial, factory farms. If you see any you think belong in this category, please let me know. Some farms charge a fee for tours, so make sure you ask first. In most cases the charges are nominal, from three to eight dollars. I also suggest you purchase goods at farms whenever possible to help compensate for the farmers' valuable time.

Farm Stands and U-Picks

These are farms that interact with the public primarily through farm stands (usually on the farm or close by) as well as U-pick operations. The most popular U-pick commodities are strawberries, blueberries, blackberries, apples, and sometimes vegetables. I included farm stands that are in good condition and sell mostly (though often not exclusively) their own goods. Because there are scores of U-pick farms across the state, I focused on those that have large amounts or varieties of offerings, a picturesque setting, and sometimes additional activities.

Apple Orchards, Apple Stands, and U-Picks

You'll find this category only in the first chapter, which covers the mountain region, as this area contains the vast majority of apple orchards. The listings highlight but aren't limited to pick-your-own orchards, as well as those that have attractive settings and views, children's activities, well-stocked sales areas, and a mix of apple varieties and products.

Farmers' Markets

There is no accurate count of farmers' markets throughout the state, but the list now easily tops 100. Most but not all of the markets included here are "producer markets," meaning the farm and craft vendors can sell only what they grow or make, and usually the vendors must live within fifty to seventy miles of the market location. My selective lists focus on markets I thought were the liveliest or most interesting.

Choose-and-Cut Christmas Trees

The state ranks second (after Oregon) in the nation in number of Christmas trees harvested, some 5 million in 2009. Most of these are Fraser firs, which are grown in the mountains. Farmers elsewhere grow pines, cedars, and other varieties well suited to warmer climates. The farms listed here sell "choose-and-cut" trees, where customers pick out their own trees and can even cut them down if they want. I focused on choose-and-cut farms offering a variety of activities, especially for children, and those that have attractive settings.

Vineyards and Wineries

North Carolina's wine industry ranks tenth in the nation in terms of size and continues to grow yearly. By the time you read this, there will likely be more than 100 wineries in our state. Most eastern vineyards grow muscadine grapes, which make a sweet wine. As you head west, you start to find European vinifera grapes, used for drier wines. But there are crossovers throughout the state. Instead of rating wines, I rated experiences. The wineries featured here are those that offer the best visiting experiences, including the setting, view, tasting room, amenities, and customer service. Most wineries have tasting fees, usually from three to eight dollars, depending on how many wines are included.

Stores

In this section you'll find a wide variety of stores that carry farm-fresh products or farm-related merchandise. These include community food cooperatives, natural-food stores, specialty food stores, gift stores, year-round farmers' markets, and even a few grocery chains.

Dining

Researching restaurants that are seriously committed to using food from local farms was a daunting task. Many dining establishments these days tout their local-food initiatives and don't follow through, while others do it as a matter of course and don't advertise it. I talked to farmers, chefs, and savvy diners to help compile this selective list. My aim was to highlight not only strict "farm-to-table" restaurants but also those that have been working with farmers long before it was trendy, as well as those that do their best under challenging conditions, such as location and size. I also sought a variety of restaurants based on price, setting, and type of food.

Lodging

Staying on a farm overnight is a wonderful way to see farm life up close and personal. Most of the accommodations listed in this category are on working farms, including goat farms, Christmas tree farms, and general family farms. Others are in farmhouses, former farms, or areas surrounded by farms and farmland.

Special Events and Activities

The yearly calendar is filled with festivals and events that celebrate our state's farm-fresh commodities, from wine to cotton to blueberries. In this category you'll also find fairs, farm tours, classes and workshops, and other special happenings.

Recipes

Each chapter ends with recipes harvested from North Carolina farmers, chefs, inn proprietors, and cookbook writers, all featuring farm-fresh ingredients and written for the home cook.

Glossary, Appendixes, Index

At the end of the book you'll find a glossary of farm terms used throughout the book, a listing of sites featured in the book by county, information on some of my favorite farm-related organizations, as well as an index that includes every listing and every recipe in the book.

Dining Price Key

$ inexpensive; most entrées under $15

$$ moderate; most entrées $15–$25

$$$ expensive; most entrées over $25

Lodging Price Key

$ inexpensive; rooms under $100

$$ moderate; rooms $100–$150

$$$ expensive; rooms $150–$200

$$$$ deluxe; rooms over $200

Talk Back to Me

With very few exceptions, I visited every farm and related business in this book. Of course, between the time I was there and the moment you pick up this book, some things will have changed. Some farmers will alter their course, businesses will move or close, and schedules will change. In addition, new and worthy farms and enterprises will open their doors.

If you visit a place you'd like to see included in the next edition, or if you see anything that needs updating, clarifying, or correcting, please let me know by writing to diane@farmfreshnorthcarolina.com. Meanwhile, I will post all updates at www.farmfreshnorthcarolina.com.

TIPS FOR VISITING FARMS

- A working farm is not Disney World. It doesn't have regular operating hours or a paved parking lot. Its mission is to provide us with food or other agricultural goods, not to entertain us.
- Most farmers' homes are on farm property. Always make an appointment before visiting, unless regular public hours are posted.
- Honor farmers' time. If you have an appointment, arrive on time. If a farm gives free or nominally priced tours, plan to purchase something while you're there.
- Farms aren't petting zoos. While you may get to collect eggs or pat an alpaca, it's best to prevent a child's (or adult's) meltdown by not promising animal interaction.
- Pack comfortable shoes that can get dirty, drinking water, hand sanitizer, sunscreen, a hat, and insect repellant. And don't forget the cooler, for holding purchases.

- Be responsible for your children, keeping them under your supervision and away from farm tools and equipment.
- Assume all fences are "hot," meaning they're electric, turned on, and will sting like the dickens if you touch them.
- Leave pets at home; they don't mix well with cows, ducks, goats, chickens, sheep, pigs, alpacas, and the farm dogs trained to protect their charges.
- Enjoy this wonderful opportunity to see where your food comes from and to visit the hard-working people who grow and raise it for you.

Mountains

From the Christmas tree farms of the High Country in the north to the apple orchards in Henderson County, the twenty-one-county mountain region is a farm-lover's paradise. Half the fun of visiting farms here is maneuvering the twisting roads in the shadow of the Appalachians. What once was a common sight — small stacks of burley tobacco drying in fields — has faded, while a new crop of sustainable produce and livestock farms has taken its place. Farm-to-table restaurants abound here, especially in the Asheville area, making the city's "Foodtopia" campaign more than a marketing slogan. One day in Asheville, we lunched on chicken salad made from Hickory Nut Gap Farm poultry. An hour later we were at Hickory Nut's farm stand, where we saw Imladris Farm jam for sale. Our next stop? Imladris Farm. Foodtopia indeed.

FARMS

Crosscreek Farm

At fifty-acre Crosscreek Farm, in a beautiful valley near the New River north of Sparta, Colette and Jonathan Nester and their two sons raise hogs, chickens, goats, and Jersey cows, selling pork, grass-fed beef, eggs, and vegetables. The house and barns date to the late 1800s, when the farm was used for dairy, tobacco, and poultry. The Nesters, who moved to this family land in 2002, have been slowly fixing up the buildings and turning this into a sustainable practice. In 2009, with grant money from Rural Advancement Foundation International, Colette started the Blue Ridge Farmers Market, a local-foods retail store for which she hopes she'll find continuous funding. Not only is Colette willing to give farm tours, she feels it's her duty. "We are so blessed to have inherited this farm. It's a gift we like to share."

2416 Nile Road, Sparta (Alleghany County), 336-372-8574, www.crosscreekfarmnc.com. Sales and tours by appointment.

Big Horse Creek Farm

Ron and Suzanne Joyner left behind science teaching and research jobs in Raleigh to work in a different sort of laboratory—the nursery where they preserve heirloom varieties of apples. Since 1997, the couple has lived on the remote, seventy-five-acre Big Horse Creek Farm in Lansing, north-west of West Jefferson. Customers put in their orders early in the year, choosing from some 300 apple varieties. The Joyners then graft the plants onto rootstock, raise them in their nursery, and ship the saplings in the fall. Because of their steep and rugged driveway, Ron and Suzanne open their farm to the public only once a year and by appointment, but you can buy their apples from spring to fall at the Ashe County Farmers' Market in West Jefferson. "All these old varieties of apples are so much better than you can get in any grocery store," Suzanne said. "The range of colors and textures is extraordinary."

Old Apple Road, Lansing (Ashe County), 336-384-1134, www.bighorsecreekfarm.com. Sales and tours by appointment.

Sundance Farm

When Barbara and Ellis Aycock started Sundance Farm in 2006, they were an anomaly—the only people to farm organically in Avery County. "We're an organic island in a sea of Christmas tree growers," Barbara said. Barbara, a retired teacher, and Ellis, a retired lawyer and former cattle and horse rancher in Burke County, have about an acre and a half in produc-tion, growing an array of vegetables and berries. They sell at a couple of farmers' markets, as well as to restaurants and off the farm. They welcome visitors to their beautifully laid-out farm to "educate people and share our love of the land," Barbara said.

61 Little Hill Lane, Newland (Avery County), 828-733-1465. Sales and tours by appointment.

At Blue Ridge Bison in Weaverville (Buncombe County), visitors can watch these shaggy beasts being fed. Photo by Selina Kok.

Blue Ridge Bison

Watching a herd of twenty-five bison traverse a rocky hillside in western North Carolina is a wondrous sight. At Blue Ridge Bison you can even see the shaggy beasts being fed by owners Leonard and Thais Wiener. Why bison? "We had to have some kind of animal to keep the vegetation low," Lenny said of pasture on their sixty acres. "They don't need a barn, so they're easier and cheaper than cattle." The meat is available directly from the farm and at small groceries and restaurants. Above the sales office is a shop of a different kind—Thais's metal studio—where she makes and sells delicate mixed-metal and gemstone jewelry.

99 Ballard Branch Road, Weaverville (Buncombe County), 828-658-3634. Sales and tours by appointment.

Double "G" Ranch

"Lance spent half his life trying to get off this road and the other half trying to get back on," Lance Graves's wife, Valerie, said with a laugh. In 2007 the couple moved from Tennessee to their twenty-seven-acre farm in Leicester, down the road from the farm Lance grew up on. Their growing collection of animals includes organically fed free-range chickens, goats, and pigs, mostly heritage breeds. In the summer, they also sell organically grown vegetables and sunflowers. "We're very kid-friendly here," said Valerie, a parent herself, who aims for visitors to get the full farm experience, including collecting eggs in an array of colors.

16 Twin Bridges Road, Leicester (Buncombe County), 828-683-6092, www.ashevilledoublegranch.com. Sales during farm-stand hours. Tours by appointment.

Earthaven Ecovillage

In 1994 a dozen people started Earthaven Ecovillage deep in the mountains southeast of Asheville. Since then it has become one of the country's best-known intentional communities. The more than fifty residents live self-sufficiently, are committed to sustainable practices, and share some common areas. A weekly tour of part of the 320 acres takes visitors throughout the community, which includes small farms, edible gardens, and a native-plant business. This one-stop farm tour option includes information on permaculture design, medicinal herbs, and off-the-grid living. The five-acre Gateway Farm located at Earthaven also conducts its own tours twice a month.

1025 Camp Elliott Road, Black Mountain (Buncombe County), 828-669-3937, www.earthaven.org. Tours by appointment.

Gladheart Farms

When the founders of Gladheart Farms bought seven acres just southeast of downtown Asheville in 2006, they saved the land from almost certain development. Since then the group has acquired more land, and the tidy, eighteen-acre certified organic farm produces a large variety of vegetables, focusing on heirloom varieties such as Asian eggplant, burgundy okra, and yard-long pole beans. Gladheart is owned and operated by about thirty-five members of Twelve Tribes, a religious community in which members give up worldly possessions, live simply, and pool work and resources.

29 Lora Lane, Asheville (Buncombe County), 828-280-7595, www.gladheartfarms.com. Farm stand May to October. Tours by appointment.

Imladris Farm

A drive to Imladris Farm takes you along country roads over hills, through valleys, and up a steep dirt drive. Once there, sixth-generation farmer Walter Harrill will give you a friendly and informative tour of his family's small-scale farm, which Harrill inherited in the 1990s. Since then, he, his wife, Wendy, and their son, Andy, have turned the farm into a sustainable showcase, focusing on berries but including rabbits and goats. From their 4,000 pounds of raspberries a year, along with a more recent addition of blueberries and blackberries, the Harrills sell about 20,000 jars of jam a year, which they make at a commercial kitchen in Candler. Make sure you allow time for a jam tasting after the tour.

45 Little Pond Road, Fairview (Buncombe County), 828-628-9377, www.imladrisfarm.com. Sales and tours by appointment.

Long Branch Environmental Education Center

In 1974 Paul Gallimore, his wife, and a few friends purchased 125 acres in the mountains northwest of Asheville. More than three and a half decades later, Paul and his wife own 1,635 acres, which he has turned into the nonprofit Long Branch Environmental Education Center. The name may sound formal but the place is laid back. Workshops and demonstrations include cider pressing using the farm's apples and sustainable building and farming practices. The farm also has become known for its U-pick offerings of blueberries, blackberries, raspberries, and apples.

278 Boyd Cove Road, Leicester (Buncombe County), 828-683-3662, www.longbrancheec.org. Sales, U-pick, and tours by appointment.

Round Mountain Creamery

"I think goats are people with four legs," said Linda Seligman, owner of Round Mountain Creamery, a dairy goat farm and Grade A goat milk processing plant. Linda, who has about 300 goats on her twenty-eight acres southeast of Asheville, started out with a few goats for fun. "After I started milking, I said, 'OK, let's make cheese.' Then I got the idea to do milk." After almost a decade of experimentation, she opened her creamery in 2008. Linda sells her fresh goat milk and several varieties of cheese in stores and from the farm. Linda's advice to humans: "Walk a path as sure-footed as a dairy goat."

2203 Old Fort Road, Black Mountain (Buncombe County), 828-669-0718, www.roundmountaincreamery.com. On-farm sales daily; call first.
Sales and tours by appointment.

Venezia Dream Farm

Starr Cash and Joe Jaworski moved here after early retirement from their high-tech jobs in California in 2000. Their forty acres off a quiet road northeast of downtown Asheville includes a mountain ridge and acres of pasture. They knew they wanted to raise animals, but hadn't decided which ones until Starr, a Kentucky native, saw a television commercial about alpacas. A month later they started their herd. They breed and sell the animals, whose fiber they send to be turned into carding and yarn. The couple also holds alpaca education workshops for would-be farmers and rents out the farm's original house to vacationers (they live up the hill). Starr offers guests farm tours and a chance to feed the alpacas, which we eagerly took her up on.

276 Jones Cove Road, Asheville (Buncombe County), 828-298-9166, www.veneziadream.com. Sales and tours by appointment and during open-farm days.

Warren Wilson College Farm

On any given day, you can see students working the land just off the road at the Warren Wilson College Farm in the mountain-flanked Swannanoa Valley. This 900-student four-year college, known for its environmental and sustainable agriculture teachings and practices, operates a 275-acre crop and livestock farm. Customers get on waiting lists to buy the farm's grass-finished beef and pasture-raised pork, sold twice yearly. Vegetables, all organically grown, are sold from May to September at the Riceville Tailgate Market, near campus. Visitors are welcome to view the farm and eat at the two spots on campus, which use farm products in season.

701 Warren Wilson Road, Swannanoa (Buncombe County), 828-771-3014, www.warren-wilson.edu/farm/.

Millstone Meadows Farm

Mark and Sara Hord had their third date at Millstone Meadows Farm in Morganton, which then grew and sold daylilies exclusively. When Mark heard the long-time owner talk about retirement, he said he might be interested in buying the beautifully landscaped farm. The next spring, the couple did just that and a month later they were married there. They've not only continued the farm, but Sara, a well-regarded chef, grows herbs and heirloom vegetables for cooking. All flowers and produce are organically grown. The Hords also raise chickens, sheep, and pork, and host weddings. In 2010, the couple started a "wine and dinner club," for which they host dinners and wine tastings at the farm.

2595 Henderson Mill Road, Morganton (Burke County), 828-433-7126, www.millstonemeadowsfarm.com. Daylily sales May to July. Other sales, tours, and dinners by appointment.

The ABCs of CSAs

More and more consumers are buying directly from local farms through a Community Supported Agriculture program (CSA), a kind of subscription service with a farmer. Throughout the state and nation, the number of CSAs has exploded over the past few years.

Some North Carolina CSAs have more than 300 members, while others have a dozen. Still others operate through markets that include goods from several farmers. Payment is upfront and for the season, which gives the farmer income to purchase plants and livestock. Food pickup is usually at the farm or at various points throughout a community. Long-running CSAs can fill up quickly, so it's best to sign up early.

CSA members usually have special benefits as well. Mother Nature willing, you're guaranteed to get certain items you sign up for, and farmers often include recipes for current offerings. The best of crops and meat are reserved for members, and small-production items, such as honey and flowers, go to members first. Farms that aren't open to the public might be open to members, regularly or on special days, so you can truly get to know your farmer.

The biggest downsides are that pickup days and times are usually locked in, and you don't have much if any control over what you're getting, like, say, a box of mostly cabbage and onions. Like the farmer, you're at the mercy of the harvest. But then, you become a more creative cook.

Yellow Branch Farm and Pottery

Shortly after moving to their sixty-acre mountain homestead near Fontana Lake in 1980, Karen Mickler and Bruce DeGroot bought a Jersey milk cow. One cow led to several more, and Yellow Branch Farm became licensed to sell cheese in 1986, becoming the state's longest-running farmstead cheese maker. Bruce makes the cheese while Karen concentrates on her beautiful stoneware pottery. The gardens and barns are open to the public, and if

Bruce is available, he's happy to explain the cheese-making process. Inside the lovely cottage that serves as Karen's pottery studio, visitors may purchase not only flavorful, buttery cheese but a handmade platter to serve it on.

136 Yellow Branch Circle, Robbinsville (Graham County), 828-479-6710, www.yellowbranch.com. Open April to October and by appointment.

Sunburst Trout Company

If you've had farm-raised rainbow trout at a North Carolina restaurant, there's a good chance it came from Sunburst Trout Company in Canton, one of the top trout producers on the East Coast. The company, run by Sally Eason and several family members, was started in 1948 by her father, Dick Jennings, as Jennings Trout Farm, the South's first commercial trout operation. It became Sunburst in 1980, when Sally and her husband, Steve, took over. Visitors are welcome to walk around the farm, where water flows from the Pisgah National Forest into the holding areas. While there is no gift shop or visitor center, customers can go to the office to buy whatever products are on hand, which might include filets, caviar, trout burgers, and Sunburst's yummy smoked tomato jam.

128 Raceway Place, Canton (Haywood County), 828-648-3010, 800-673-3051, www.sunbursttrout.com. Open year-round. Tours by appointment.

Wildcat Ridge Farm

Wildcat Ridge Farm focuses on a few specialties. It sells prize-winning U-pick/they-cut peonies and dahlias, along with possibly the South's largest selection of container-grown fig trees, in Mediterranean, Italian, and southern varieties. U-pick raspberries and blackberries are available, too. The figs are a passion of Canton farmer Ricardo Fernandez, who by night is chef and owner, with his wife, Suzanne, of Lomo Grill in downtown Waynesville. He's also the creator of Chef Ricardo's Authentic Appalachian Tomato Sauces, sold in stores statewide. The farm's stunning setting along the Pigeon River is another reason to visit this out-of-the-way spot.

3553 Panther Creek Road, Clyde (Haywood County), 828-627-6751, www.wildcatridgefarm.com. Open May to October.

Carl Sandburg Home

The state's most visited goats reside in Flat Rock, near Hendersonville, at Connemara Farms Dairy. While the herd is now tended to by National

Park Service staff and volunteers, it once was the pride and joy of Lilian Sandburg, wife of famed poet and author Carl Sandburg. Since 1974, the couple's home and farm have been the focal points of the Carl Sandburg Home National Historic Site. The Sandburgs kept the existing farm name, Connemara, when they moved to Flat Rock in 1945 in order for Lilian to expand the dairy business she had started in Michigan. She managed a herd of 200 goats until 1966 and was internationally known for her expertise in genetics and breeding. Today, many of the fifteen or so goats there are direct descendants of Lilian's. The well-maintained barn, milk house, and other buildings are open for touring (cheese-making demonstrations are held on summer weekends). Park-goers also are allowed to enter the grazing fields to visit with the famed goats.

Little River Road off Highway 225, Flat Rock (Henderson County), 828-693-4178, www.nps.gov/carl. Open year-round.

Doubletree Farm

Most of the quilts found painted on the sides of barns are traditional block designs. Not so at Doubletree Farm, where four squares of horse heads more closely resemble an Andy Warhol print than a bedspread. "Yeah, ours is a little different," noted Cathy Bennett. She and her husband, Andy, also do other things a little differently on their thirty-two acres. While they grow produce, which they sell at market, they're known locally for using draft horses for their sustainable logging operation and for making sorghum syrup, similar to molasses. During most of September and October, Doubletree extracts the syrup from its own and other farmers' sorghum cane. One or two horses, walking in a circle, turn the mill that squeezes out the juice. The Bennetts usually schedule a few days when visitors can see natural horse power in action.

835 Cargile Branch Road, Marshall (Madison County), 828-689-3812. Sales and tours by appointment.

Eagle Feather Organic Farm

Robert Eidus, owner of Eagle Feather Organic Farm and North Carolina Ginseng and Goldenseal Company, grows herbs, plucks wild ginseng and goldenseal from his wooded property, runs workshops, does consulting, and hosts a plant show on community-access television. He became hooked in 2000 after attending a workshop with expert herbalist Jeanine Davis at the North Carolina Mountain Horticultural Crops Research and

Extension Center. "The plants speak to me," said Robert, whose twenty-eight acres near Marshall include rustic workshop space and sheltered camping.

300 Indigo Bunting Lane, Marshall (Madison County), 828-649-3536, www.ncgoldenseal.com. Sales, tours, and workshops by appointment. Lodging $

East Fork Farm

For a dozen years after moving from Hendersonville to Madison County, Dawn and Stephen Robertson led a fairly solitary farming existence, raising lamb and poultry to sell and growing produce for themselves and their two daughters. Not until 2008 did they go public, selling at two farmers' markets, to Greenlife Grocery, several restaurants, and from the farm. They welcome visitors to their lovely forty-acre spread, which travels up a mountainside. The farm is best known for its lamb, grass-fed on rotational grazing pastures. East Fork also keeps laying hens, some rabbits for meat, and pond-raised trout. In 2009, Dawn and Stephen built an A-frame cabin just beyond their home for vacation rentals.

215 Meadow Branch Road, Marshall (Madison County), 828-206-3276, www.eastforkfarm.net. Sales and tours by appointment. Lodging $$

Spinning Spider Creamery

To Chris Owen, it's all about the kids—goat kids and human ones. The idea for this family business started when one of her three sons raised a goat as a 4-H project. One goat led to another, and now the children are part of this award-winning artisanal cheese business in Madison County, in operation since 1996. The farm keeps around eighty to 100 goats, some on the family's five acres at home in the mountains near Mars Hill and others on sixty acres nearby. Chris welcomes visitors to the farm to purchase cheese, but she doesn't always have time to chat. "I had to tell one group, 'I've just gotten into some gouda; you have about an hour until I can stop stirring.'" The creamery includes a cheese cave and a small retail space. Chris gives some cheese-making classes and holds visitation days throughout the year, keeping all the kids busy.

4717 East Fork Road, Marshall (Madison County), 828-689-5508, www.spinningspidercreamery.com. Sales during farm-store hours and open-farm days. Tours by appointment.

Chris Owen makes several varieties of goat cheese at Spinning Spider Creamery in Marshall (Madison County). Photo by Diane Daniel.

Wake Robin Farm Breads

Using organic flours from Lindley Mills in Alamance County, Steve Bardwell and Gail Lunsford make several varieties of breads and sweets on their century farm outside of Marshall. Many of their other ingredients, including eggs and molasses, are local as well. The couple has been a strong supporter of local farms and food, even before they opened their bakery in 2000. They helped start the farmers' market in Marshall and later organized the annual Asheville Artisan Bread Festival. From their farm, which is under a forest management plan to reintroduce the American chestnut, they host occasional cooking classes taught by area chefs and give tours.

472 Teague Road, Marshall (Madison County), 828-683-2902.
Classes and tours by appointment.

Meadowbrook Nursery/We-Du Natives

At Meadowbrook Nursery/We-Du Natives, half of this attractive nursery's twenty-eight acres is reserved for mountain natives—rhododendron, mountain laurel, and azaleas. Overall, this enchanted garden tucked along a curving back road fifteen minutes south of Marion specializes in native plants of the Southeast. "We're better known in New York and Boston than we are here," said horticulturist Jamie Oxley, who has owned Meadowbrook with his wife, Merri, since 2002. In operation since 1981, it does a brisk mail order business throughout the country, and also is a destination nursery for green-thumbed travelers. Tours are given only to groups and master gardeners, but anyone is welcome to browse and shop. The first Saturday in October is "Plantober Fest," with public tours and lectures throughout the day.

2055 Polly Spout Road, Marion (McDowell County), 828-738-8300,
www.we-du.com. Open year-round. Tours by appointment.

Peaceful Valley Farm

The fifth and sixth generations of the McEntire family live in the old farmhouse at Peaceful Valley Farm, set in the magnificent Crooked Creek Valley in Old Fort. Owners John and Karen McEntire have turned the original farmstead into an educational area. Visitors come to see the hayloft, barn, and corn crib, where feed is ground for the chickens, rabbits, and goats, and a popular festival is held here every fall. When we visited, brothers John and Jerrill (who owns part of the 124 acres) fired up a 1950s tractor to start the cane mill, once powered by mules. In went sorghum cane and out came syrup. We left with a jarful and a bag of stone-ground grits from Peaceful Valley's heirloom corn.

1200 Pine Cove Road, Old Fort (McDowell County), 828-668-7411.
Fall festival in October. Sales and tours by appointment.

Harrell Hill Farms

Harrell Hill Farms is a double century farm, having been in the Harrell family for more than 200 years. After a few decades away, Doug Harrell returned to the farm in 1990 with his wife, Barbara. When she saw their early 1800s farmhouse she asked her husband, "Are we going to raze it or burn it?" Instead, they restored it, and the Harrells are happy to give visitors tours of their historic home on 200 acres. While the farm has had a

Jerrill McEntire of Peaceful Valley Farm in Old Fort (McDowell County) squeezes juice from sorghum cane to make syrup. Photo by Diane Daniel.

long-standing choose-and-cut Fraser fir operation as well as wholesale tree sales, the Harrells in 2009 decided to expand their agritourism offerings. Doug built a shelter for his restored 1919 cane mill, where visitors can watch sorghum syrup (similar to molasses) made from the farm's sorghum crop. He also turned a dairy barn into a shop for sales of trees, vegetables, and meat from the family's grass-fed Limousin beef cattle, and he gives hayrides and farm tours.

467 Byrd Road, Bakersville (Mitchell County), 828-688-9188, www.harrellhillfarms.com. Farm-stand hours September to Christmas. Otherwise, sales and tours by appointment.

Mighty Forces in the Local Food and Farm Movement

North Carolinians are lucky to have two strong and thriving organizations concerned about the livelihood of independent farms.

The Carolina Farm Stewardship Association (CFSA; www.carolinafarmstewards.org), based in Pittsboro, near Chapel Hill, was founded in 1979 by a group of farmers and consumers to foster the growth of organic food in the Carolinas. Indeed, this membership organization has helped define and expand the state's sustainable agriculture movement. The group counts as members more than 1,100 farmers, processors, gardeners, businesses, and individuals in North Carolina and upstate South Carolina. With only a small staff, its ambitious projects include an online local food and farm directory, two weekend farm tours that draw thousands, a seed bank, an organic flour project, and political involvement in the regional food system. Every year at alternating locations around the state, CFSA puts on the popular Sustainable Agriculture Conference, filled with farm tours, dozens of workshops and classes, farm-fresh meals, and national speakers.

If you've ever been west of Statesville, chances are you've seen a "Local Food: Thousands of Miles Fresher" bumper sticker. The sticker is one of the most visible signs of the Appalachian Sustainable Agriculture Project (ASAP; www.asapconnections.org), an Asheville-based nonprofit formed in 1999 to support farmers and rural communities in western North Carolina and the southern Appalachians. ASAP, like CFSA, works to create a regional community-based food system. The dozen or so staff members carry out an impressive number of projects, including a top-notch Local Food Campaign. Through free printed and online Local Food Guides, ASAP lists hundreds of farms, markets, wineries, farm-friendly restaurants, lodging options, and more. It also started an "Appalachian Grown" food labeling program, operates a farmers' market and a farm tour, matches chefs and farmers, and organizes an in-depth Farm to School program. In 2009, ASAP unveiled its Farm to Hospital pilot program, helping ten area hospitals incorporate locally grown foods.

Mountain Farm Museum

Along with the natural attractions at Great Smoky Mountains National Park are educational ones, including the wonderfully arranged Mountain Farm Museum, on the Cherokee side of the park at the Oconaluftee Visitors Center. Given the park's 9 million annual visitors, this is probably the most visited educational farmstead in the state. Situated in an open field, it includes a furnished two-story log house from the 1840s, a barn, pig pens, corn cribs, apple house, small orchard, and outbuildings. The restored buildings have been gathered from around the Smokies. Crops planted in the gardens include heritage vegetables and broomcorn, the source of broomstraw, used to make brooms.

150 Highway 441 North, Cherokee (Swain County),
828-497-1940, www.nps.gov/grsm/. Open year-round.

Apple Hill Farm

The logo for Apple Hill Farm in Valle Crucis, near Boone, shows an apple flanked by a donkey and a goat, with an alpaca on top. Owner Lee Rankin planned to have only alpacas on the forty-three acres she bought in 2001. But after an animal she thinks was a mountain lion killed several of her alpacas, she got llamas to guard the alpacas, donkeys to bray warnings, and goats as a sacrificial meal. Since those additions, life on this lovely farm in the mountains has been peaceful but busy. In 2009 Lee opened the farm seasonally on Saturdays for tours, and she arranges private visits other days. Her setting and setup are superb. Even the barn is beautiful enough to live in. The tour includes a look at the naturally grown produce garden, berries, apple orchard, and, of course, the alpacas. A lovely shop in the barn carries goods made from alpaca fleece, goat-milk soap, and other handmade products. "What we really like is teaching people about the animals," Lee said. "If a child learns that animals have feelings, then I've done my job."

400 Apple Hill Road, Banner Elk (Watauga County), 828-963-1662,
www.applehillfarmnc.com. Sales and tours May to October or by appointment.

Faith Mountain Farm

We were a few feet from the front door at Faith Mountain Farm when the aroma of fresh-baked goodies reached us. Items for sale here are written on a chalkboard next to the front door. Shannon Wilkes ushered us into a kitchen whose shelves were stacked with loaves of organic bread, honey buns, cookies, and organic granola, sold here and at the Watauga County Farmers' Market. Her husband, James, a beekeeper, tends to twelve colonies, so honey often is for sale as well. In 2005 the couple and their seven children moved from Boone to this sixty-five-acre farm in Creston, an hour north. "Before, in the city, we were creating artificial things for the kids to do for work; now there's plenty," Shannon said with a laugh.

489 Big Laurel Road, Creston (Watauga County), 336-385-3510,
www.faithmtnfarm.com. Sales on drop-in basis. Tours by appointment.

The Farm at Mollies Branch

Diane Price will tailor tours at The Farm at Mollies Branch to all ages and interests, but there's one thing she insists on for everyone. "You must get a llama kiss," she told us. Don't worry, it's on the cheek, and it's really a nuzzle. We quite enjoyed ours. Diane, who bought the fifteen-acre farm in 1996, keeps her chickens in the barn, where they're guarded by the five to ten llamas. Other highlights at Mollies Branch, which is just around the corner from Todd, north of Boone, are a look at Price's micro-hydroelectric system, which powers half her house and all the barn, and a walk up the holler to the shiitake mushroom logs.

156 Timberlakes Drive, Todd (Watauga County), 828-264-1337.
Tours by appointment.

Goodnight Family Sustainable Development Teaching and Research Farm

Appalachian State University students in sustainable agriculture learn small-scale farming and organic production techniques at their own farm outside of Boone. The ten-acre Goodnight Family Sustainable Development Teaching and Research Farm was created in 2001. Highlights include a modified passive solar greenhouse, medicinal herb garden, permaculture and biointensive gardens, cold frames for extending the growing season, an apple orchard, and a re-created wetland. A self-guided walking tour is available to the public in printed form or online, and there's even a hiking trail around the perimeter.

Dutch Creek Road (a quarter mile off 194 South toward Banner Elk),
Valle Crucis (Watauga County), 828-262-7268, http://susdev.appstate.edu.

Peak Cut Flower Farm

After twelve years of operating Shady Grove Gardens and Nursery in Zionville, Susan Wright and Brent Cochran decided to expand by buying a farm to grow their own flowers for cutting. Since 1998, when they bought seventy acres in Creston, an hour north of Boone, the couple has gradually started to open up the remote mountainside Peak Cut Flower Farm to visitors. In summer and early fall they offer weekly hour-long hayride tours, where Brent describes farm operations and the natural history. Since 2005, the couple has hosted weddings on a grassy knoll with views of North Carolina, Tennessee, and Virginia. Most recently, they became one of the few farms on the North Carolina Birding Trail, and they are developing a trail system through the woods.

2278 Peak Road, Creston (Watauga County), 828-297-4098, www.shady-grove-gardens.com. Tours by appointment.

Apple Brandy Beef

Three years after graduating from North Carolina State University in 2004 with a degree in agribusiness management, third-generation farmer Seth Church led the transformation of the family's cattle business. He turned the 300-acre farm into a "birth-to-beef" direct-market operation, under the name Apple Brandy Beef. Seth acknowledges that most of his fellow no-hormone, no-antibiotic livestock farmers raise grass-fed cattle, whereas Apple Brandy's are corn fed. "Our customers prefer the corn fed," he said. "Our market is toward those who want the prime cuts." Within two years, Earth Fare stores and a list of high-end restaurants were carrying Apple Brandy Beef. The meat also is sold off the farm, near North Wilkesboro, and online. Seth welcomes visitors to his green rolling hills. "It's my obligation to the industry to show that the cattle are well cared for, to the benefit of the animals, the environment, and our customers."

3746 Mountain View Road, North Wilkesboro (Wilkes County), 336-696-2721, www.applebrandybeef.com. Sales and tours by appointment.

Tumbling Shoals Farm

Some people follow their parents into farming or inherit farmland and start working it. Others, like Shiloh Avery, take the textbook approach. She studied sustainable agriculture at Pittsboro's Central Carolina Community College in Pittsboro and worked at the Triangle's Peregrine Farm, a state leader in sustainable farming. She and her husband, Jason Roehrig, then spent three years looking for land. The young couple found the perfect spot in a scenic valley in Millers Creek, north of Wilkesboro. On four acres, called Tumbling Shoals Farm after the creek that cuts through the valley, they are "trying to grow everything," Avery said, which means more than 100 varieties of produce and flowers. She expects the farm to be certified organic by 2012. Avery and Roehrig sell at markets, through a CSA, and twice weekly from the farm.

841 Sand Ridge Road, Millers Creek (Wilkes County), 336-452-2920, www.tumblingshoalsfarm.com. Sales during farm-stand hours. Tours by appointment.

EnergyXchange's Project Branch Out

Nurseries are a huge business in North Carolina, but many of them are run like factories, selling nonlocal varieties in nonnatural ways. Project Branch Out at EnergyXchange in rural Burnsville is working to change that. Its mission is to propagate rare and native azaleas and rhododendrons of western North Carolina from seed and tissue culture, reintroducing some varieties of native deciduous plants that have become rare in the nursery business. The plants can then be transplanted to nursery fields, thereby helping to protect natural flora populations. While sales here are mostly wholesale (thirty-plant minimum), some retail sales are offered. But you don't have to purchase anything to visit the greenhouse. The nonprofit EnergyXchange, fueled by methane gas created on what once was a two-county landfill, also is one of the nation's model energy recovery projects, and it houses artists' studios for glass blowers and potters.

66 EnergyXchange Drive, Burnsville (Yancey County), 828-675-5541, www.energyxchange.org. Open year-round. Tours by appointment.

Maple Creek Farm

Though Maple Creek Farm grows vegetables and raises pigs, its claim to fame can be found in its name. The 106-acre farm northwest of Burnsville has the country's southernmost commercial sugar bush, the name for a grove of sugar maple trees. Since 2007, farm manager and forester Richard Sanders has been tapping sugar maples for syrup every February, starting with a few taps and not much syrup and growing to around 500 taps and close to 100 gallons of syrup. "Up north they tap with buckets, but here it's way too hilly and rocky," said Richard. To solve that problem, he has run three miles of tubing through the grove to tap the trees, whose

Visitors to Mountain Farm near Burnsville (Yancey County) can view the Black Mountain range from the fields of lavender. Photo by Diane Daniel.

sap flows by gravity. Sanders also grows sorghum cane and makes sorghum syrup every fall.

1641 Lickskillet Road, Burnsville (Yancey County), 828-682-0297, www.maplecreekfarm.net. Open-farm days in September and February. Tours by appointment.

Mountain Farm

Mountain Farm in Celo, south of Burnsville, is one of western North Carolina's most appealing and successful agritourism enterprises. Marilyn and Jerry Cade, who both work in medicine, have lived here since 1974. In the late 1990s they opened the farm to the public. It now features two acres of naturally grown lavender and a lavender labyrinth, dairy goats, U-pick blueberries, a lovely little shop carrying the farm's lavender products and goat-milk soap, and, most recently, a homemade ice cream and coffee bar. Visitors are invited to view the goats, in a barn near the shop, and stroll through the fields of lavender up the farm's highest peak, with wide views of the Black Mountain range. The Cades also rent out Blueberry Cottage, a 1930s farmhouse across the street. Guests can visit the farm anytime during their stay and help with farm chores. The farm's early summer Lavender Festival and the September Fall Festival each attract hundreds of visitors.

125 Copperhead Bend, Burnsville (Yancey County), 828-675-4856, www.mountainfarm.net. Summer and fall events. Farm stand open year-round. Lodging $$

Wellspring Farm

Though she was a corporate executive for many years, Elke Amende-Spirakis always knew it wasn't her life's work. "I grew up doing fiber arts, like spinning and knitting," Elke said. Later, living in New York state, she started attending county fairs. "I saw a llama, and the rest is history." By 1999, when she and her husband left their townhouse in New York for their forty-five-acre Wellspring Farm outside of Burnsville, "we had four angora rabbits at home and four llamas boarded." Now, at Wellspring Farm, everyone can be together, including llamas, sheep, and angora goats and rabbits. Elke recently added a shop for her fiber demonstrations and products. Tours include time with the llamas and spinning demonstrations, and Elke holds occasional open-farm days during shearing season.

166 Wellspring Lane, Burnsville (Yancey County), 828-682-0458, www.wellspringfarm.com. Sales and tours by appointment.

FARM STANDS AND U-PICKS

Old Orchard Creek

After former Raleigh residents Walter Clark and Johnny Burleson purchased Christmas tree farmer Dale Shepherd's 1890s home place near Lansing in 2003, they continued many traditions. One was to renovate the historic house and another was to protect the land by placing conservation easements on the eighty-eight-acre farm. But the one the public is most familiar with is the continuation of the farm's U-pick berry operation. The 3,200 heritage blueberry bushes at what is now called Old Orchard Creek attract a steady stream of visitors each summer to pick berries. Walter and Johnny have made the farm, which includes an apple and peach orchard, even more inviting by adding covered picnic tables that look out onto the 5,000-foot-high Pond Mountain.

410 Swansie Shepherd Road, Lansing (Ashe County), 336-384-2774, www.oldorchardcreek.com. Open July to August.

When They Say Organic . . .

At farmers' markets and elsewhere you'll find some farmers saying their crops are "organic," or "organically grown," while others say they are "certified organic." In 2002, the U.S. Department of Agriculture (USDA) imposed national standards, called the National Organic Program.

To be deemed "organic," farms selling $5,000 or more worth of goods labeled organic must keep compliance records and pay to be certified yearly through an accredited certifying agent. Even farms selling less than $5,000 worth must follow the same rules if they call those products organic, but they don't need to be certified. No farm of any size is allowed to use the term "organically grown" unless it follows the program guidelines. Many farmers ignore that edict, sometimes out of ignorance and sometimes out of contempt for what they say is the government's co-opting of the term *organic*.

What makes food organic? Organic meat, poultry, eggs, and dairy products come from animals that are fed organic feed and given no antibiotics or growth hormones. Organic food is produced without using most conventional pesticides, petroleum-based fertilizers, or genetically engineered products.

The organic program guidelines ban most but not all synthetic chemicals and pesticides, and they allow many naturally produced ones, which means organic farms are not necessarily "pesticide-free" or "chemical-free."

Some products display the label "Certified Naturally Grown." This is a grassroots alternative to the USDA program, involving what the group says are the same standards yet with less paperwork and at a lower cost. The farmer-run program, based in New York, is meant primarily for small farmers distributing through local channels.

Still other farms that for whatever reason do not wish to be certified organic may be nonetheless free of chemicals and pesticides, following biodynamic principles or their own strict standards.

Dogwood Hills Farm

U-pick options abound at this lovely rural ten-acre spread, which young farmers David Nielsen and Jenny Zoppo started managing in 2008. All produce is organically grown. Cherry picking starts in June, followed by blueberries through July, and thornless blackberries and Concord grapes in August. Apple production continues from late June to mid-November.

369 Ox Creek Road, Weaverville (Buncombe County), 828-645-6286. Open June to November.

Farside Farms

Farside Farms products, including eggs, and hormone and antibiotic-free pork, chicken, and beef, are well represented on grocery store shelves in western North Carolina. For one-stop shopping, you can visit Farside's own farm market in the Woodfin community north of Asheville. While the setting along a busy highway isn't a draw, the money you'll save on the farm's products compared to supermarket prices might be. Much of the in-season produce is the farm's, though the store does supplement from near and far.

83 Weaverville Road, Asheville (Buncombe County), 828-683-3255. Open year-round.

Flying Cloud Farm

While most of the naturally grown produce from Flying Cloud Farm's eleven acres is sold at farmers' markets and through a CSA, the farm also operates a small and very tidy self-service roadside stand. Not only does the two-sided cart sport covered containers for the items, the produce is put on ice. Owned by young farmers Annie Louise and Isaiah Perkinson, Flying Cloud also sells a small amount of beef and pork and offers U-pick strawberries in May and June and blueberries in June and July.

1860 Charlotte Highway, Fairview (Buncombe County), 828-628-3348, www.flyingcloudfarm.net. Open June to October.

Hickory Nut Gap Farm

You've likely seen or tasted Hickory Nut Gap Farm products, as the 600-acre farm's beef, lamb, pork, chicken, and eggs are distributed widely to stores and restaurants. The fifth-generation family farmers practice sustainable agriculture on their sixty acres of livestock pasture. For one-stop shopping and farm visits, stop by the beautifully situated farm store in

Fairview, where all Hickory Nut products are sold, as well as foods and gifts from other area farms. In September and October, the store adds apples (some organically grown), and for kids, there are barnyard animals, a pumpkin patch, and a corn maze. When it's time for refreshments, try a yummy "cidersicle."

57 Sugar Hollow Road, Fairview (Buncombe County), 828-628-1027, www.hickorynutgapfarm.com. Open year-round.

Stoney Hollow Farm

Scott Boxberger and family moved from Charlotte to the far western reaches of the state in 1998 to start farming. A decade later, he and his wife, Stephanie, their seven children, and his parents were firmly entrenched at Stoney Hollow Farm, set off a quiet country road with mountain views. The Boxbergers run a U-pick and they-pick produce and berry farm, which includes a wide range of vegetables, fruit, and berries, grown on about ten of their 200 acres. "Things are really maturing now," said Scott, who also keeps ten beehives. From their self-service farm stand, they sell not only produce and honey but a mouth-watering array of jams and baked goods, all made by Stephanie in her adjoining commercial kitchen. If all the picking and shopping wears you out, a beautiful rental cabin is available for overnight stays.

944 Ollies Creek Road, Robbinsville (Graham County), 828-735-2983, www.stoneyhollowfarm.net. U-pick open April to November. Store open year-round. Lodging $$

The Ten Acre Garden

Danny Barrett loves when adults bring children to The Ten Acre Garden in Canton. "There aren't a lot of kids who get to dig in the dirt anymore," he said. Barrett's ten-acre farm, surrounded by mountains, started as a small garden with his children. He opened the stand in an attractive barn-like building in 2005. Though he worked for thirty-six years at the paper mill in Canton, retiring in 2008, "I've lived and breathed the farm for twenty years," he said. "It's a century farm, and my ancestors settled this valley." Barrett has several acres of U-pick strawberries, raspberries, and blueberries, along with many types of vegetables, some heirloom varieties, as well as herbs and flowers.

276 Chambers Farm Lane, Canton (Haywood County), 828-235-9667. Open May to September.

Edmundson Produce Farm Market

In 2007, after years of growing and selling produce from their 200-acre farm near Hendersonville, the Edmundson family opened its own store, Edmundson Produce Farm Market. In the summer and fall, the majority of produce is from the family. Grandfather, father, and son are all master bee-keepers and sell their honey, while mom tends to the commercial kitchen housed at the store. From there she makes baked goods, jams, and salsas to sell. Other regional farms supply some produce, eggs, and beef.

3771 Brevard Road, Horse Shoe (Henderson County), 828-891-3230.
Open year-round Monday through Saturday.

Deal Family Farm

The Deal family has been farming commercially in the mountains of Macon County since 1951. With some eighty acres of produce in production on two dozen tracts of land around Franklin, Deal Family Farm supplies a little bit of everything to wholesalers as well as stocking its two produce stands, one east and one west of town. Third-generation brothers run the business now. In October, the Deals operate a five-acre corn maze, offer hayrides, hold a weekend-long harvest festival, and open up their large pumpkin patch for picking.

4402 Murphy Road (Highway 64), Franklin (Macon County), 828-524-5151;
and 96 Deal Farms Circle, Franklin, 828-524-3774. Open April to December.

J. W. Mitchell Farms

Though J. W. Mitchell Farms outside of Franklin has been open only since 2003, farmer John Mitchell has been working the soil since he was a youngster in Florida. When Hurricane Andrew severely damaged that state in 1992, John was still farming, and he and his wife, Dorothy, were insurance agents. "I had to shut down the farm and work on paying claims," he said. In 2000 they moved to land they owned in the North Carolina mountains. Within a few years John was growing dozens of varieties of vegetables and flowers on thirty-five acres. John operates a farm stand as well as the only full-service fully U-pick farm we've encountered in the state. "It's extremely gratifying for Dorothy and me to see grandparents bring their grandchildren," he said. "A lot of the kids have never seen the plants their food comes from."

405 Bradley Creek Road, Franklin (Macon County), 828-349-2725,
www.jwmitchellfarms.com. Open July to October.

Okra is among the dozens of vegetables available for U-pick at J. W. Mitchell Farms outside of Franklin (Macon County). Photo by Selina Kok.

Zimmerman's Berry Farm

In 2000 Pam and Billy Zimmerman turned to U-pick berries to help save their former tobacco farm, set deep in the mountains in central Madison County. It helps that they love to eat them, too. "Before growing them, I never had all the berries I could eat," Pam said. Picking season on their eighty-five acres (three are for the berries) begins in mid-June with black raspberries and ends around mid-August with red raspberries and blueberries. In between are blackberries and wineberries. From her tiny farm stand, Pam also makes jams, jellies, and syrups from the berries, which she finally gets her fill of every summer.

2260 Revere Road, Marshall (Madison County), 828-656-2056, www.zimmermansberryfarm.com. Open June to August.

OakMoon Farm and Creamery

"I saw *Heidi* when I was a kid and was totally taken by the goats on the mountain," said Cynthia Sharpe, who runs OakMoon Farm and Creamery in Bakersville with her husband, Dwain Swing. The farmstead creamery has been open since 2007, but Cynthia has been making cheese since the 1980s. The tiny sales area is in the entryway to OakMoon's basement creamery in town, and it runs on the honor system. The farm, home to about seventy goats, is a mile up the road. We couldn't resist buying a chunk of Clementine, a natural-rind aged raw goat-milk cheese. Cynthia advised us it would get tastier with age, but we couldn't wait that long.

57 Highway 226 North, Bakersville (Mitchell County), 828-688-4683, www.freewebs.com/oakmoonfarm/. Open year-round.

The Manna Cabanna

Saluda residents should consider themselves lucky. This cute-as-heck mountain town, with fewer than 600 residents, can shop for organic and sustainable produce much of the week from spring to fall at The Manna Cabanna. Owner Carol Lynn Jackson, an enthusiastic proponent of healthy eating, opened her small, cheery stand in 2008 and has been drawing visitors and locals since. She also started both a summer and winter CSA using products from all her farmers. "Everything here is from within sixty miles and was harvested just hours ago," Carol Lynn said.

105 East Main Street, Saluda (Polk County), 828-817-2308. Open May to October.

Farmers' Favorites

BRAND OF CLOTHING: Carhartt

LIVESTOCK GUARD DOG: Great Pyrenees

READING MATERIAL: Seed catalogs

LUNCH: Tomato sandwich with Duke's mayonnaise

DRINK: Unpasteurized cow or goat milk

SOCIAL TIME: Farmers' market

TIME TO NAP: Saturday afternoon, after the farmers' market

VACATION SPOT: Anywhere, as long as they can get back to
the farm within an hour

BEST FRIEND AND WORST ENEMY: Mother Nature

Darnell Farms

Darnell Farms has been selling a variety of produce in Bryson City for more than thirty years. Its market, next to the scenic Tuckaseigee River, is the perfect place to hold its many weekend events, which include a springtime strawberry festival, a summer tomato celebration, and a fall corn maze. The store supplements its produce from near and far, but the views of the Smokies are 100 percent local.

2300 Governors Island Road, Bryson City (Swain County), 828-488-2376.
Open April to November.

Queens Produce and Berry Farm

Even though he sold Queens Produce and Berry Farm to younger folks in 2003, Dee Queen greeted us when we arrived for a July visit. "Mr. D," born in 1924 and a fixture on the farm, still helps oversee the fourteen acres in production. We were happy to have him give us a deluxe tour, past rows of beans, beets, potatoes, cucumbers, eggplants, peppers, rhubarb, and more. Everything sold at the farm stand, inside the barn, is grown on the farm, which is situated in a valley a few miles north of Brevard. Along with produce, Queens has a thriving U-pick business with blackberries, raspberries, blueberries, and rows of flowers. "We sell sunflowers by the bucketful," Mr. D said.

858 Davidson River Road, Pisgah Forest (Transylvania County), 828-884-5121, www.queensberryfarm.com. Open spring to fall.

Four Winds Berry Farm

Marsha Bryant could see that her career in the furniture-manufacturing industry wasn't going to get her to retirement. Plan B was pursuing a passion for growing things. After getting an associate's degree in horticulture, she took the leap and in 2007 planted fifteen acres of blueberries, raspberries, and blackberries on the beautifully situated forty-acre Four Winds Berry Farm in Moravian Falls. "Part of the challenge was trying to plant on the least steep hills," she said. It's those hills that make this such a lovely place to visit, with a wide view of the Brushy Mountains from the top. The U-pick operation, which started in 2009, is made even more inviting by the tiny farm store housed in a restored apple house.

7493 Highway 16 South, Moravian Falls (Wilkes County), 336-667-6769, 828-612-6902. Open June to September.

APPLE ORCHARDS, APPLE STANDS, AND U-PICKS

Deal Orchards

Since 1939, Deal Orchards has been growing fruit on its hundred acres of orchards north of Taylorsville, about forty-five minutes northwest of Statesville in the Brushy Mountains. Starting in July, the orchard opens a simple shop in its packing house to sell peaches, nectarines, Asian pears, several varieties of apples, and pumpkins in the fall. The store stays open until the apples in cold storage run out, usually at the end of February.

7400 Highway 16, Taylorsville (Alexander County), 828-632-2304. Open July to February.

Apple Hill Orchard and Cider Mill

In a rural area of rolling hills in Enola, south of Morganton, Gayle and Harley Prewitt operate Apple Hill Orchard and Cider Mill, the region's only remaining commercial apple orchard, operating since 1957. Twenty-five acres of trees and a dozen varieties of apples provide U-pick and store customers plenty of fall fruit. On Saturdays from early September to mid-November, scenic tractor-drawn wagon rides take children and adults through the orchard, traveling along Appletree Lane, where trees were first planted in 1930. Apple Hill is also known for fresh-pressed pasteurized cider, served warm or cold from its country store.

2075 Pleasant Hill Avenue, Morganton (Burke County), 828-437-1224, www.applehillorchard.com. Open August to December.

J. H. Stepp Farm's Hillcrest Orchard

The Stepp family operated a U-pick before there was such a term. When they took over the farm in 1967, they continued the practice started by the former owner to let neighbors pick up apples late in the season. So it's not surprising that the Stepps' Hillcrest Orchard became the county's first official U-pick in 1972. You can view the orchard's progression through photographs in the retail shop. Outdoors, though, is where the real action is. On most days you're likely to encounter groups of kids and adults here to pick and purchase, some of whom have been returning for decades.

221 Stepp Orchard Drive, Hendersonville (Henderson County), 828-685-9083, www.steppapples.com. Open August to November.

Justus Orchard

The Justus family has been cultivating apples and produce for more than a century. In the shop at Justus Orchard, their twenty-five-acre farm in the hills, you'll find cider, homemade jams and jellies, and an assortment of whatever vegetables are in season, most of them grown on the farm. You can buy fresh-picked apples or pick your own. Pack a lunch to eat at the covered picnic area, and make sure to leave room for the orchard's divine apple chips, which are dehydrated on the premises. In the summer, Justus has U-pick blackberries.

187 Garren Road, Hendersonville (Henderson County), 828-685-8033, 828-685-8167, www.justusorchard.com. Open July to November.

Lyda Farms

"Did you get your apple?" a worker at Lyda Farms farm stand asked as she handed us what all visitors receive, a complimentary apple, in this case a crunchy red delicious. Members of the Lyda family have been farming in North Carolina since the 1700s, and in the twentieth century Rome Lyda was a leader in developing the state's apple industry. The family also grows and sells produce. Children know Lyda for its locally famous insect, "Juicy the Giant Apple Bug," a colorful worm gracing the front lawn.

3465 Chimney Rock Road, Hendersonville (Henderson County), 828-685-3459, www.lydafarms.com. Open June to November (apples arrive in August).

Piney Mountain Orchards Produce

Owners Sara and Wade Edney's roots run deep. Not only are they third-generation fruit growers, they're descendants of Samuel Edney, who in the late 1700s settled Edneyville, where most of Hendersonville area's orchards are located. Fruit from the family farm is sold at their Piney Mountain Orchards store, the Edneys' seasonal shop north of the city. What the highway site loses in setting it makes up for in seasonal offerings. "We grow almost everything we sell," Sara says. "Apples, cherries, peaches, blackberries." Also impressive is the store's collection of beautifully decorated gourds—all grown and painted by Sara.

3290 Asheville Highway, Hendersonville (Henderson County), 828-692-4800. Open April to November. Gourds also sold at www.simplygourdgeouslive.com.

Sky Top Orchard

If there were an apple-picking heaven, this would be it. First, there's the winding road up Mount McAlpine, with glorious views along the way, matched only by the view some 3,000 feet up. Then there are the Butler family's fifty acres of orchards gently sloping downward. On weekends the pick-your-own fun includes hayrides, a barnyard, and other children's activities. Picnic tables and a play area invite customers to linger. At the farm store you can buy some of the orchard's twenty-five varieties of apples, as well as grapes, Asian pears, and peaches in season. And don't overlook the snacks—dried apples, cider slushies, and the best cider doughnuts you'll ever taste, even in heaven.

1193 Pinnacle Mountain Road, Zirconia (Henderson County), 828-692-7930, www.skytoporchard.com. Open August to December.

Windy Ridge Organic Farms

Despite North Carolina's being the state that produces the sixth-greatest amount of apples in the country, harvesting 165 million pounds a year, only one large-scale orchard grows apples organically. A decade later, not only is Windy Ridge Organic Farms still the only certified organic apple orchard in the state, it's the only one in the entire Southeast. The reason? Growing organic apples is more difficult, and expensive, because of diseases, insects, heat, and humidity. But for farmer Anthony Owens, the extra work pays off. Historically a wholesale-only operation, Windy Ridge will open to U-pickers in 2012. "I got so many requests I finally decided to do it," said Anthony. In 2011, he will open the orchard's first retail roadside stand.

4041 Chimney Rock Highway, Hendersonville (Henderson County), 828-712-1919, www.windyridgeorganic.com. Seasonal.

The Historic Orchard at Altapass

The state's most famous orchard sits right along the Blue Ridge Parkway outside of Spruce Pine. The Historic Orchard at Altapass was planted in 1908 by the Clinchfield Railroad and went through many lives and owners. When the 280 acres of land was offered for sale in 1995, local real estate agent Kit Trubey bought the orchard, and then she; her brother, Bill Carson; and his family went about preserving it. Today Altapass is one of the finest agritourism sites in the state, offering live music, storytelling, hayrides, a butterfly preserve, a gift shop (online as well) and, in the fall, U-pick and prepicked apples from its 3,000 trees, many of them old-time varieties. Many of the events and activities are supported by the nonprofit Altapass Foundation, which the families created to promote preservation of the orchard and this storied Appalachian community.

1025 Orchard Road, Milepost 328.3 on Blue Ridge Parkway, Spruce Pine (McDowell County), 888-765-9531, 828-765-9531, www.altapassorchard.com. Open May to November.

Apple Mill

Not only does the Apple Mill have its own orchard, the family business has produced fruit butters, jams, preserves, and ciders for decades. You'd be hard pressed to go into any mountain store and not spot its products. At its well-stocked outlet store in Saluda, you'll find all of Apple Mill's offerings as well as natural, unpasteurized cider, a rarity these days. If you're lucky enough to arrive on a cooking day, from the inside of the store you can see through a window into the kitchen and watch fresh fruit butter brewing in a 200-gallon kettle.

1345 Ozone Drive, Saluda (Polk County), 828-749-9136, www.ncapplemill.com. Open year-round.

FARMERS' MARKETS

Ashe County Farmers' Market

During peak times, about forty farmers and artisans line up under the covered pavilion at the popular Ashe County Farmers' Market in West Jefferson. This long-running market attracts regulars and tourists visiting this popular mountain city. Special events are held throughout the season, including cooking and baking demos. The market also has contributed to the draw of downtown, a block away and usually hopping on Saturday mornings. Though the regular market closes in October, four holiday craft markets operate through December.

10 North Second Avenue, West Jefferson (Ashe County), www.ashefarmersmarket.com. Held Saturday mornings April to October.

Asheville City Market

Downtown Asheville is ringed by almost a dozen farmers' markets, with the most recent, and one of the most popular, debuting in 2008. Area farmers, the city of Asheville, and the folks at the Appalachian Sustainable Agriculture Project can be credited for getting the popular Asheville City Market off the ground. Shortly after its spring opening with forty vendors, the market took off and now operates with close to seventy-five farmers and artisans, most from within sixty miles of Asheville. Children's activities, market tours, live music, and chef demos using goods from the market give local-food shoppers yet another reason to venture downtown.

161 South Charlotte Street (Public Works Building), Asheville (Buncombe County), 828-236-1282, www.asapconnections.org/citymarket.html. Held Saturday mornings April to December.

Black Mountain Tailgate Market

A few blocks from the center of downtown Black Mountain lies the lively Black Mountain Tailgate Market. Set on the shaded grounds of First Baptist Church, the market has plenty of room for kids to play and picnickers to spread out their blankets. The market has about twenty vendors selling mostly organic and sustainably grown produce, plants, flowers, and herbs, as well as food items and crafts. This also is the Saturday market for Foothills Family Farms, a progressive farm collective out of Old Fort representing some fifteen farmers.

130 Montreat Road, Black Mountain (Buncombe County),
www.blackmountaintailgatemarket.org. Held Saturday mornings May to October.

Haywood's Historic Farmers' Market and Waynesville Tailgate Market

In 2009, a group of farmers in the Waynesville area formed Haywood's Historic Farmers' Market. They hold it only a block away from the long-running Waynesville Tailgate Market, and the two run on almost identical schedules. The tailgate market comprises the old-timers, who sell straight from the back of their pickup trucks. The newer and, when we visited, larger farmers' market is made up mostly of younger farmers who sell from the shelter of pop-up tents. That market also has more organic and sustainable produce, as well as farm-based products, such as soaps and candles.

Haywood's Historic Farmers' Market, 250 Pigeon Street; Waynesville (Haywood County), www.waynesvillefarmersmarket.com. Waynesville Tailgate Market, 171 Legion Drive, Waynesville. Both held Wednesday and Saturday mornings May to October.

Flat Rock Tailgate Market

Like the village it's situated in, the Flat Rock Tailgate Market is small, colorful, and lively. It was formed in 2007, when a group of local women and retailers interested in food and cooking decided they needed a market close by instead of having to drive to one in another town. Produce, meat, and other products are all sustainably grown. The downtown boutiques and eateries of Little Rainbow Row have added to the market's great success. As for the founders, no more driving is required.

2720 Greenville Highway, Flat Rock (Henderson County), 828-697-7719,
www.flatrockonline.com. Held Thursday afternoons May to October.

Henderson County Curb Market and Henderson County Tailgate Market

For more than thirty years, shoppers in downtown Hendersonville have had two farmers' markets to choose from, both featuring products grown or made in Henderson County. The impressive Henderson County Curb Market is the old-timey and older one. Run by a co-op of nearly 100 families, the indoor market has been in operation since 1924 and even owns

Ask Your Extension Agent

Have questions about gardening, farming, goat tending, bee-keeping, cooking, nutrition, and more? Chances are your county extension office can help. The North Carolina Cooperative Extension Service became official in 1914 (as the Agricultural Extension Service), though its roots were laid earlier as part of a social movement to help working-class people gain practical education to improve their lives. Its mission today is largely unchanged, "to help individuals, families and communities put research-based knowledge to work for economic prosperity, environmental stewardship, and an improved quality of life."

North Carolina's extension service is based at N.C. State University in Raleigh and is a cooperative effort with North Carolina A & T State University in Greensboro and the U.S. Department of Agriculture. Extension specialists and researchers at both schools support the field staff, who work in each of the state's 100 counties to bring knowledge from the experts to the public.

The 4-H (head, heart, hands, health) program for youth, which celebrated its 100th birthday in 2009, is run through the extension service. So is the Master Gardener program, started in 1979, which every year educates thousands of citizens, who pass along their wisdom to the rest of us.

its own building. Produce accounts for a small percentage of offerings, which include jams and jellies, baked goods, and folk crafts. The outdoor Henderson County Tailgate Market, open since 1979, is the place to go for produce from organic and sustainable farms. It has about forty food and craft vendors. Since the markets are only a few blocks apart, it's easy—and recommended—to visit both.

Curb Market, 221 North Church Street, Hendersonville (Henderson County), 828-692-8012, www.curbmarket.com. Open Saturday mornings year-round, Tuesday and Thursday mornings April to December. Tailgate Market, 100 North King Street, Hendersonville, 828-697-4891. Held Saturday mornings April to October.

Jackson County Farmers' Market

In Sylva, population 3,500, the Jackson County Farmers' Market positively buzzes, and we're not just talking about the educational beehive brought in by the gals at Balltown Bee Farm in Bryson City. We counted nearly forty vendors on a toasty July morning in the Smoky Mountains. Among the vendors were several sustainable produce farmers, as well as makers of crafts and farm products, including jams, soaps, and candles. Market members also organize a county farm tour and garden walk every July, an impressive feat for a relatively small group.

31 Allen Street (next to Bridge Park), Sylva (Jackson County), 800-962-1911. Held Saturday mornings May to October.

Madison County Farmers and Artisans Market

With about twenty-five farmer and craft vendors, the Madison County Farmers and Artisans Market in Mars Hill is the largest in the county. Formed in 1998, it's set on the lovely small-town campus of Mars Hill College, a block from the tiny downtown. Typical offerings include local vegetables, fruits, and herbs, as well as baked goods, jams, and goat cheese. Local musicians perform during the peak season.

Off Dormitory Drive near Pittman Dining Hall, Mars Hill (Madison County), www.marshillmarket.org. Held Saturday mornings April to October.

Cherokee Farmer's Tailgate Market

Vendors at the Cherokee Farmer's Tailgate Market in Cherokee sell the usual varieties of produce, but you'll also find locally traditional heirloom vegetables, such as flour corn and "lazy housewife beans." Only enrolled families of the Eastern Band of Cherokee Indians can sell items here, which means the market stays quite small, usually from five to eight vendors. The market is about a mile from downtown next to the Cherokee Reservation extension service—just look for the farmers' market sign printed in English and Cherokee.

876 Acquoni Road, Cherokee (Swain County), 828-554-6931. Held Friday mornings July to October.

Transylvania County Tailgate Market

The Transylvania County Tailgate Market in downtown Brevard has been going on for "twenty-something years," said organic farmer Dianne Hill, but it remained fairly small until 2005, when farmers started marketing the market. "Now, on Saturday mornings it's wild in there," she said. Saturdays in the summer attract around forty farm and craft vendors, and the weekday market draws a handful of farmers. Offerings include produce, meat, cheese, flowers, baked goods, and jams.

212 South Gaston Street (next to library), Brevard (Transylvania County),
828-884-9483, 828-883-3700. Held Saturday, Tuesday, and Thursday mornings
April to December.

Watauga County Farmers' Market

The High Country's largest market is Watauga County Farmers' Market in Boone. In operation since 1974, the lively market draws hundreds of visitors and eighty to ninety farm, food, and craft vendors during peak season. Farmers must be from Watauga or bordering counties. Along with the usual offerings, we saw a few market firsts for us—a wheatgrass grower and a raw-food bakery, called Hold the Heat. Turning up the heat here are hot sauce makers Fire from the Mountain, who grow their own peppers. While you're at the Watauga market, consider a stroll through the adjacent Daniel Boone Native Gardens, which has an impressive collection of native plants.

591 Horn in West Drive, Boone (Watauga County), 828-355-4918,
www.wataugacountyfarmersmarket.org. Held Saturday mornings
May to October and Wednesday mornings June to September.

CHOOSE-AND-CUT CHRISTMAS TREES

Joe Edwards Christmas Tree Farm

Melia Edwards's father added a choose-and-cut business to his wholesale operation by popular demand. "My dad started this farm in the mid-1960s and people would drive by and ask, 'Can I come back at Christmas for a tree?' So he started the first choose-and-cut in the state." Now Melia and her husband, Joe, run the Joe Edwards Christmas Tree Farm in Sparta, which features fifty acres of Fraser firs and Colorado spruce and one of the best views around—a panorama of the Blue Ridge Parkway. In 2007 the couple added a two-level art gallery, largely to sell Joe's art, which includes oil paintings, watercolors, and wire-wrapped jewelry, along with photographs taken by both husband and wife.

2081 Pine Swamp Road, Sparta (Alleghany County), 336-372-1711,
www.blueridgefineart.com. Open late November to December.
Gallery open March to December.

About 2 million Christmas trees a year are harvested in Ashe County, making it the largest tree producer in the state. Photo by Selina Kok.

Papa Goat's Tree Farm

Dennis Bell is the farmer and old goat at Papa Goat's Tree Farm in Sparta. "My grandkids always call me Papa Goat, so I had to name it that," said Dennis. Judging from the activities at his family's festive choose-and-cut Christmas tree farm, Dennis clearly knows what puts a smile on a child's face. A bevy of barnyard animals (goats, donkeys, and ducks) near the red-and-green gift and snack shop, complete with miniature red silo, will captivate the young ones' attention. Adding to the merriment are the hayrides that take adults and children to the twenty acres of nearby fields to pick out their trees. And if all this activity leaves you too tired to drive home, Papa Goat's also has two cabin rentals.

687 Three Creeks Lane, Sparta (Alleghany County), 336-372-3055, 704-618-7360. Open late November to December. Lodging $$

Lil' Grandfather Mountain Christmas Tree Farm

We had to check out Lil' Grandfather Mountain Christmas Tree Farm during the week, though it's open only on weekends. Still, we figured we could see the mountain setting, which we'd heard was breathtaking. Then, it turned out the fog was so thick, we couldn't see five feet ahead of us. But enough locals endorsed Lil' Grandfather that we do, too. The Deal family started the farm in 1985 with two acres of trees as a hobby. Now it has sixty acres, more than 100,000 trees, and operates a choose-and-cut and wholesale business. Along with Fraser firs, it grows Colorado blue spruce and white pine. Choose-and-cut weekends are filled with activities, including hayrides, food and craft sales, and sometimes pony rides. Just try to go on a clear day.

15371 Highway 18 South, Laurel Springs (Ashe County), 336-359-8817, www.lilgrand.com. Open late November to December.

Mistletoe Meadows

While you probably don't have a place in your house for a tree similar to the one Mistletoe Meadows provided to the White House in 2007—an eighteen-foot Fraser fir—it's nice to know that any tree from here is in esteemed company. This Laurel Springs farm, owned and operated by Joe Freeman, also sells trees and wreaths for the average homeowner. Joe started the farm in 1988 at the age of twenty-seven, many years younger that most tree farmers. Now he grows trees for wholesale and retail sale on 130 acres in Ashe County and in Virginia. The choose-and-cut operation in Laurel Springs has spectacular mountain views. Joe credits his wife, Linda, for the name Mistletoe Meadows, a suggestion that we hope was sealed with a kiss.

583 Burnt Hill Road, Laurel Springs (Ashe County), 336-982-9754, www.mistletoemeadows.com. Open late November to December.

Shady Rest Tree Farm

While the family-run Shady Rest Tree Farm, just off the Blue Ridge Parkway, has weekend activities for kids, including hayrides, hot chocolate, lollipops, and coloring books, what sets it apart is its keen focus on silvaculture. For this you can credit Tim Miller, a thirty-year tree farmer who is always working to grow healthier trees with fewer pesticides. While he cultivates Fraser firs, like everyone in the mountains, he also has Nordmann, Concolor, and Canaan firs, as well as Colorado blue spruces and white pines for garlands. If you're interested in learning more about the science of Christmas trees, this is a great place to visit.

287 Trading Post Road, Glendale Springs (Ashe County),
336-877-1908, 336-982-2031, www.shadyresttreefarm.com.
Open late November to December. Shady Rest also ships trees.

Wayland's Nursery

We arrived at Wayland's Nursery, southeast of West Jefferson, on a frigid, blustery day in December, not wanting to get out of the car. We'd driven up a nearly vertical paved road, then a snaking dirt one. At the top, where the nursery and Wayland and Lynn Cox's home are located, awaited a dramatic 360-degree view. Precut Fraser firs and wreaths lined the drive, though customers also can go into the fields and choose their own tree. Also at the top is Blueberry Hill Gift Shop, filled with Christmas and blueberry-themed merchandise in the gift shop—all made by Lynn. In the summer, the Coxes sell blueberry plants and U-pick berries when they're available. Lynn said visitors love walking into the shop to say, "We're looking for a thrill on Blueberry Hill." We thought the drive up was thrill enough.

1003 Round Knob Ridge Road, West Jefferson (Ashe County), 336-246-7729.
Open late November to December. U-pick berries open July to August.

West End Choose and Cut and West End Wreaths

West End Choose and Cut grows Fraser firs, but the real star at this farm is West End Wreaths, which has won several state and national awards. What makes the wreaths so special? "Probably our wreath makers," said co-owner Scott Ballard. "We want wreaths to have dimension, thickness." In the loft of its seventy-five-year-old barn, West End makes 1,000 wreaths a day during full production. Each takes about six to seven minutes to make, using special tables holding a clamping tool operated by a foot pedal. Yearly, West End sells about 20,000 wreaths wholesale and 2,000 by retail mail order. Keep a fresh wreath away from direct sunlight, Scott advised, "and they'll stay green even when they're dry."

2152 Beaver Creek School Road, West Jefferson (Ashe County), 336-846-7300,
877-207-1661, www.westendwreaths.com. Open late November to December.

O (Organic) Christmas Tree

With North Carolina being the second-leading producer of Christmas trees in the country, you'd think there would be at least one sizeable organic tree farm. There's not. Rogue Harbor Farm in Marshall does sells wreaths from its certified organic Fraser firs, but it has very few and sometimes no trees for sale. A few other mountain farmers sell trees billed as "pesticide free." Rogue Harbor owners Aubrey and Linda Raper, who ship their handmade wreaths all over the country, say that growing attractive organic trees is possible, but it takes more work and a longer growing time.

In the Triangle, forestry students at Duke University in 2008 started planting Leyland cypress and eastern red cedar saplings for their student-run organic Duke Forestry Christmas Tree Farm. They expect the first trees to be ready in 2012 or 2013, with tabletop sizes coming sooner.

And up in Ashe County, the top tree-producing county in the state, North Carolina extension agent Bryan Davis in 2008 started testing an acre of organic Fraser firs for research. Bryan is an expert in integrated pest management (IPM), which has significantly reduced the amount of pesticides used in the tree industry over the past fifteen years. The underlying philosophy is to exhaust all natural methods before turning to pesticides and then to use as little as possible. A significant part of IPM is the planting of clover and other ground covers, which helps suppress weeds, reduces erosion, and attracts wildlife. Bryan estimated that more than half the state's tree growers now practice IPM.

Elk River Evergreens

Many choose-and-cut farms give away hot chocolate and cider, but when you need carbs, head to Elk River Evergreens for freshly made popcorn. Along with the usual holiday crafts, the shop carries a great selection of homemade wreaths, including special shapes such as crosses, stars, and candy canes. Elk River has been a tree wholesaler since 1975 and grows more than 400,000 trees on twelve farms. In 2002 it added this choose-and-cut retail operation near the Tennessee border. Co-owner Gary

Edwards, when he was an Avery County commissioner years earlier, helped Fraser fir growers reduce their pesticide use.

121 Brooks Shell Road, Elk Park (Avery County), 828-387-7695, www.elkriverevergreens.com. Open late November to December.

Evergreen Ridge Choose and Cut Farm

Evergreen Ridge Choose and Cut Farm sits between Newland and Banner Elk, and, as the reindeer flies, is close to both. But driving there, through Hickory Nut Gap, is a twisting, turning adventure that lasts several miles. At 4,600 feet in elevation, Evergreen is one of the highest choose-and-cut farms in the state, with majestic views of the surrounding mountains. The farm, which started in 1980, has been operated by the Pitman family since 1992. To entice people to make the trek, the farm offers weekend hayrides, cookies and cider, and farm animals, though the view alone is worth the trip.

299 Tower Road, Newland (Avery County), 828-733-9606, www.evergreenridgechristmastrees.com. Open late November to December.

Franklin Tree Farm

Ask local experts which mountain farmers use the least pesticides and Dean Franklin's name always comes up. You can see the results at Franklin Tree Farm in Linville Falls, just off the Blue Ridge Parkway. The trees look more natural, and the clover and grass blanketing the ground around them is higher and wilder. The choose-and-cut action centers around a small red shack, while the Franklins' century-old farmhouse sits on a hill above the trees. Also for sale here are lovely seasonal sachets made from fresh fir needles.

10645 Linville Falls Highway, Linville Falls (Avery County), 828-765-2518. Open late November to December.

Sugar Plum Farms

Ask folks to recommend a Christmas tree farm that's fun for kids and everyone mentions Sugar Plum Farms in Plumtree. Along with the natural beauty of the mountain setting, an hour southwest of Boone, and its acres of Fraser firs, Sugar Plum is as festive as it gets. Tree and wreath sales and a gift shop are housed in a row of bright red buildings, all lavishly decorated. Ducks float (or skate) on the pond, and kids can roast marshmallows around a fire pit. We dropped by while a group of schoolchildren was meeting Santa in his little house, complete with oversized mailbox. Later we caught up with co-owner Helen Pitts, decked out as Mrs. Claus, rushing to ready a wreath order for UPS pickup. Indeed, Santa's work is never done.

1263 Isaacs Branch Road, Plumtree (Avery County), 888-257-0019, 828-765-0019, www.sugarplumfarms.com. Open late November to December. Tours by appointment.

Sandy Hollar Farms

We happened to visit picturesque Sandy Hollar Farms choose-and-cut farm the week co-owner June Hawkins' expanded and gussied up gift shop opened. "I'm just tickled to death to be able to show it to you," said June, who fills the shop not only with Christmastime gifts but also scarves, shawls, and hats made from fiber she spins herself from the farms' sheep, angora goats, and llamas. She and her husband, Curtis, planted their first trees on their 200 acres in Leicester's Big Sandy Mush community in 1970, and now their son, Dale, tends to them. When they started, folks said they couldn't grow Fraser firs at their altitude, 2,850 feet, but the farm's thriving evergreens have proved otherwise.

63 Sandy Hollar Lane, Leicester (Buncombe County), 828-683-3645.
Shop open July to December. Antique tractor show in October.
Tree sales late November to December.

Ty-Lyn Plantation

Ty-Lyn Plantation is a bit off the beaten path, but that doesn't keep more than 1,500 customers a year from coming here for their Christmas trees. Some of that is likely due to the gorgeous mountain setting on fifty-four acres near Lake Glenville, but probably most can be credited to the care that owners John and Joni Wavra put into the shopping experience (including discounts for returning customers). Instead of the usual tractor-pulled hayrides, the Wavras shuttle delighted kids and adults in a restored 1950 Chevy truck or 1951 Ford fire truck. Both are powered partly by environmentally friendly hydrogen. On weekends, made-to-order hamburgers and hot dogs are for sale. Although tucked away, the farm is easy to find—just follow the roadside Santas painted by artist John.

971 Lloyd Hooper Road, Cullowhee (Jackson County), 828-743-3899, www.tylynplantation.com. Open late November to December.

J & D Tree Farms

The best time to visit J & D Tree Farms is at the end of the day, making sure you're at the property's 4,010-foot peak at sunset. From the top you can see Elk Knob State Park and Snake Mountain. "People tell me that although they came to get a tree, the view alone was worth the drive," said Jim Bryan, who has operated the thirty-five-acre farm with his wife, Dorothy, for ten years. Bryan, an arborist, stands alone in offering the option of natural unsheared Fraser firs. Not only is he into trees, Bryan's a wind geek and hopes to some day build wind-powered rental cabins.

576 Bryan Hollow Road, Boone (Watauga County), 828-262-1845, 828-773-7925, www.jdtreefarm.com. Open late November to December.

Swinging Bridge Farm

Chuck Lieberman likes to joke that he's Watauga County's only Jewish Christmas tree farmer. At Swinging Bridge Farm, which he and his wife, Eleanor, started in 1980, you can find an unusual assortment of plants: Fraser fir trees, rhododendrons, blueberries, and citrus trees. Chuck's previous work was in the Florida citrus trade, and in Boone he has built an orangery, a sort of greenhouse for growing citrus year-round. He also sells lemon and lime trees. "Every manor house once had an orangery, and every nobleman ate oranges," said Chuck, who leads tours on his sixteen acres. And, yes, there really is a swinging bridge, ingeniously built by Chuck in 2005 to cross a stream between two blueberry fields. A tollhouse on one side asks for a crossing fee of ten cents. "It's a joke," he said, "but people keep leaving dimes."

711 Old Glade Road, Deep Gap (Watauga County), 828-264-5738, www.swingingbridgefarm.com. Citrus tree sales year-round. Plant sales spring to fall. Tree sales late November to December. Tours by appointment.

What Fir!

Unlike many of the mountains' Christmas tree farmers, What Fir! owners Nathaniel and Kirby Maram don't have farming in their blood. In 1997 the couple wanted to leave city life in Charlotte and found the perfect piece of land on Rich Mountain above Boone. It came with forty acres, four ponds, and thousands of Fraser firs. They decided to give trees a chance, and by 2005 the Marams had won a statewide award for their sustainable farming practices. What Fir!, one of the area's highest tree farms, at 4,000 feet, also offers a wonderful choose-and-cut experience, from the drive up the mountain to the gift shop along the pond. Sometimes alpacas are on hand for petting, though we're not sure if they're up to pulling Santa's sleigh.

330 Wolf Ridge Trail, Boone (Watauga County), 828-297-4646, www.whatfirtreefarm.com. Open late November to December.

VINEYARDS AND WINERIES

Banner Elk Winery

As the first winery in the High Country, Banner Elk Winery is paving the way for wine making in the state's higher counties. It was started by vintner Richard Wolfe and his wife, Dede Walton, on a twenty-acre working blueberry farm (still open for U-pick berries in August). Richard also helped start and originally led Appalachian State University's mountain and steep-slope viticulture program. The attractive tasting room includes a fireplace inside and covered porch outside overlooking a pond and pastures. Across the drive is Blueberry Villa, a luxury bed and breakfast the couple run. Guests receive a complimentary tasting and tour at the winery, and the rooms, some with fireplaces, are named for grape varieties. The porch overlooks blueberry fields and, of course, vineyards.

60 Deer Run Lane, Banner Elk (Avery County), 828-898-9090, www.bannerelkwinery.com. Lodging $$$

Biltmore Winery

While Biltmore Winery is said to be the country's most visited winery, drawing about 1 million people a year, we can assume that the Biltmore Estate is the primary attraction here. Still, the winery does a brisk business. And while only about 15 to 20 percent of grapes the Biltmore uses are from North Carolina, that small percentage equals a huge supply of Carolina grapes, around 250 tons (208,000 bottles). Most are grown in nearby Polk County. "What's great for us is we can actually go there and check on the grapes," said winemaker Sharon Fenchak. About ten vineyards grow grapes exclusively for Biltmore, which bottles its North Carolina wines under the Biltmore Estate label. The Biltmore also grows ninety-four acres of its own grapes, making the family-owned business a major force in the state's wine industry. Near the wine-tasting area and shop are other agricultural attractions. Antler Hill Farm, which houses barn animals, gives visitors an idea of what it was like to live and work on the estate in the

1890s. And nearby is the four-acre vegetable garden harvested for use in the Biltmore's restaurants.

1 Approach Road, Asheville (Buncombe County), 800-411-3812, 828-225-1333, www.biltmore.com. Biltmore Estate admission required.

Calaboose Cellars

You can't describe Calaboose Cellars without using the word *cute*. The country's smallest complete freestanding winery is housed in a 300-square-foot building that was once the first jailhouse in the small mountain town of Andrews. Out back is a small deck overlooking the "vineyards," which in this case means a few rows of French-American hybrid grapes. Another acre of grapes are grown nearby, and the rest come from other North Carolina vineyards. As gimmicky as Calaboose may sound, owners Eric and Judy Carlson are serious about their wine. Since opening in 2008, the tiny winery has won some big state awards.

565 Aquone Road, Andrews (Cherokee County), 828-321-2006, www.calaboosecellars.com. Closed January and February.

South Creek Vineyards and Winery

A big, beautiful red barn is the first thing you see from the gravel drive leading to South Creek Vineyards and Winery in Nebo, east of Marion. Just around the curve, you reach two sloped acres of vines leading up to the lovely 1902 two-story farmhouse that houses the tasting room and small gift shop. South Creek, open since 2007, is the retirement project of Frank Boldon, who owns the winery with his wife, Debra. "I'm living out a dream I've had for a long time," Frank said. "My objective is to show that vinifera grapes can be grown in McDowell County. I want other folks to grow grapes and open wineries." Customers are welcome to picnic on the six-table deck overlooking the vineyards. Bring plenty to eat, because you'll want to linger.

2240 South Creek Road, Nebo (McDowell County), 828-652-5729, www.southcreekwinery.com. Closed January to March.

Green Creek Winery

Alvin Pack is a sixth-generation Polk County native who left in the 1960s and returned in 2000 to start Green Creek Winery with his wife, Loretta. Alvin's brother, Melvin, owns and oversees the ten-acre vineyard that supplies the grapes. Alvin and Loretta lived in California wine country for several years, and Alvin modeled Green Creek's tasting room, open since 2005, after one of his favorites in Sonoma County. Unlike most Tar Heel rooms, this one has floor-to-ceiling windows, and the patio meets the vines. The nearby hitching post is for the many customers who arrive by horseback, as two trail systems in this equestrian-crazed county pass the winery. The winery's claim to fame is Chardonnay Rosso, a "red chardonnay," released in 2007, made from a mix of white chardonnay and red chambourcin grapes.

413 Gilbert Road, Columbus (Polk County), 828-863-2182, www.greencreekwinery.com.

Rockhouse Vineyards

We were lucky enough to arrive at Rockhouse Vineyards near Tryon on a late August day when winemaker Lee Griffin and crew were crushing viognier grapes. Rockhouse is the hobby turned business of Lee and his wife, Marsha Cassedy, who bought their 200-acre farm with the intention of planting grapes. It didn't take long for the small winery, open since 1998, to gather awards. The setting and tasting room are compelling as well. Lee credits Marsha with convincing him to keep the remnants of farm life standing, including two crumbling brick chimneys, a tobacco barn, and a sweet-potato house. The gorgeous "rockhouse," which holds the tasting room, was built in the 1950s by the former owners, a German couple. The exterior is made of embedded rocks, shells, glass, and fossils they collected in their travels.

1525 Turner Road, Tryon (Polk County), 828-863-2784, www.rockhousevineyards.com. Closed January and February.

STORES

Spin a Yarn

With the Ashe County Farmers' Market closed in the off-season, Nancy Hoffman of Foxfire Holler Farm needed an outlet to sell meat and fiber from her family's sheep, so she opened Spin a Yarn in 2006. From a cheery space in historic downtown West Jefferson, she and her daughter, Christine, fulfill many a yarn lover's dream by stocking a great supply of fibers and yarn both locally grown and commercially produced. Coolers in the front of the store hold lamb, pork, and beef from their farm and others

There Aren't Too Many Cooks in This Kitchen

At some point, successful food entrepreneurs and farmers who make "value-added" products from their harvest need to get out of their home kitchen and into a commercial one. To help them along, food business incubators with kitchen space are cropping up across the state. Chances are, they've all turned to Blue Ridge Food Ventures (www.blueridgefoodventures.com) for inspiration and information.

That business incubator, founded in 2005 near Asheville by the regional economic development group Advantage West, became the state's first for farmers and is among the largest in the Southeast. Since then, it has assisted more than 150 farmers and product makers, in everything from cooking space to packaging and putting North Carolina products into stores near and far. Those include Bamboo Ladies bamboo shoot pickles, Fire from the Mountains hot pepper sauces, and Imladris Farm berry jams.

Every fall Blue Ridge Food Ventures opens up its facility, at the Enka campus of Asheville-Buncombe Technical Community College, during Food Ventures Marketplace, where the public is invited to sample and purchase products.

nearby, as well as locally produced goat cheese and farm-fresh eggs. Spin a Yarn also sells handcrafts, needlework, holds classes in fiber arts, and offers occasional "yarn tastings."

8A South Jefferson Avenue, West Jefferson (Ashe County), 336-844-4771, www.spinayarnnc.com.

Black Mountain Farmers Market

Harry and Elaine Hamil don't have far to travel to stock their store, Black Mountain Farmers Market (not to be confused with the town's tailgate market, which Harry helped start). They just go to their own farm and nursery, where the couple raises naturally grown heirloom vegetables, including more than 100 varieties of tomatoes, twenty of basil, thirty of peppers, and many types of berries. Both are longtime Black Mountain residents who have been touting "eat local" way before it was a trend. What the Hamils don't grow they buy from other local farmers. The store also carries local meat, cheese, and body care and gift items. If you arrive at the store during raspberry season, you can treat yourself to a berry from the bushes grown at the entrance. We honored the "one berry per family" sign, but it wasn't easy.

151 South Ridgeway Avenue, Black Mountain (Buncombe County), 828-664-0060.

Earth Fare

What started in 1975 as Dinner for the Earth, Asheville's first natural food store, eighteen years later became Earth Fare, a leader among organic grocers. While small compared to Whole Foods, Earth Fare has long been highly regarded for supporting sustainable farmers. The chain now has about twenty stores in five southern states, with seven in North Carolina. Each store works with regional farmers to include local products on the shelves and labels them accordingly. The produce areas in stores opened since 2009 have been designed to look more like a farmers' market, with signs educating customers about the products.

Flagship store at 66 Westgate Parkway, Asheville (Buncombe County), 828-253-7656, www.earthfare.com.

French Broad Food Co-op

The French Broad Food Co-op started as a buying club in 1975 in members' homes. Today it's one of the state's most popular and liveliest co-ops, thanks in part to its location just on the edge of vibrant downtown Asheville. The store goes out of its way to bring in local products and marks them with stickers and signs. Those include produce, honey, eggs, meat, cheese, and trout, along with locally made bath and body products. On Saturday mornings from spring through fall, a farmers' market operates in the parking lot, featuring about twenty vendors who are committed to sustainable farming.

90 Biltmore Avenue, Asheville (Buncombe County), 828-255-7650, www.frenchbroadfood.coop.

Greenlife Grocery

Before you even enter Greenlife Grocery, you can read about its relationships with regional farmers. Lining its front windows are placards showing farmers and describing their products, from produce to meat and cheese.

When Chattanooga-based Greenlife opened a 20,000-square-foot full-service grocery in Asheville in 2004, it already had healthy competition from several local co-ops, farmers' markets, and Asheville-based Earth Fare stores. But Greenlife quickly established itself by hosting a tailgate market, which now runs three days a week, showcasing organic and local producers, and enticing diners with healthy to-go meals.

70 Merrimon Avenue, Asheville (Buncombe County), 828-254-5440, www.greenlifegrocery.com.

Trout Lily Market

How refreshing to find a convenience store that carries healthy food. From its start as a local food co-op, Trout Lily Market in Fairview, south of Asheville, has retained its healthy roots. Former co-op member Susan Bost bought the store in 2000 and stocks it with produce, meat, and cheese from regional and local farmers, some from just down the street, including Hickory Nut Gap meats and Flying Cloud Farm produce. The shop also carries the usual natural-foods store fare, from granola to shampoo, and a great selection of beer from regional breweries.

1297 Charlotte Highway, Fairview (Buncombe County), 828-628-0402.

Hendersonville Community Co-op

Just southeast of downtown you'll find the Hendersonville Community Co-op, which is home to an organic and natural food grocery store and the Blue Mountain Deli, where you can get home-cooked dishes to go or eat in. In 2009 the co-op started a summertime weekly farmers' market on Monday afternoons. During harvest months, the market's produce is about 65 percent local, all either certified organic or grown using organic practices.

715 South Grove Street, Hendersonville (Henderson County), 828-693-0505, www.hendersonville.coop.

Spring Ridge Creamery

In the early 1990s, dairy farmer Jim Moore decided to change his business plan and sell milk directly to customers instead of wholesale. He thought he'd throw in a little ice cream on the side. Turned out the sweet stuff became the main event at Spring Ridge Creamery in Otto, which sits along the Little Tennessee River and busy Highway 441, two miles from the Georgia line. Vacationers and second-home owners from the Atlanta area make it a ritual to stop here on their way to or from the mountains. Spring Ridge sells two dozen flavors of ice cream, along with milk, buttermilk, butter, cheese, and even farmstead cottage cheese, something we haven't seen elsewhere.

11856 Georgia Road (Highway 441), Otto (Macon County), 828-369-2958.

Good Stuff

Madison County's best collection of farm-fresh offerings can be found at Good Stuff, a funky little shop in downtown Marshall. This combo grocery store, teahouse, and gallery features organic and local foods, along with local art. The day we visited, we spotted bottles of mead concocted at Fox Hill Meadery, powder puffs made from fiber-producing goat and sheep at Philosophy Farm, jams and salsa cooked up at Sunset Valley Farm, and artisanal cheese produced from the goats at Three Graces Dairy. It's clear that owners Amy Gillespie and Jon Curtis, who previously operated an art gallery in Hot Springs, are committed to supporting local farmers and artists.

133 South Main Street, Marshall (Madison County), 828-649-9711, www.goodstuffgrocery.com.

Beneficial Foods Natural Market

Not long after it opened, residents of Adawehi Institute and Healing Center planted a garden and started preparing fresh food. About sixty people live at the 100-acre intentional community in Columbus, and many more come for workshops on such topics as self-actualization and wellness, or for sessions with alternative healers. In 2008 Adawehi greatly expanded its food services, opening the 2,800-square-foot Beneficial Foods Natural Market. Here the public can shop for vegetables, fruit, and berries from the center's biodynamic garden, herbs grown around the store, as well as dishes prepared at Beneficial's commercial kitchen. Among the creative offerings are homemade protein bars and garlic powder from the garden. Visitors are welcome to stroll the wooded walking path, visit the garden (tours by appointment), or stay over at Adawehi's bed and breakfast.

93 Adawehi Lane, Columbus (Polk County), 828-894-5260, www.jackiewoods.org. Lodging $$

The Cottage Craftsman

Among all the traditional gift shops in all the tourist-attracting towns we've visited, The Cottage Craftsman in Bryson City stands out for its local food offerings. Owner Debbie Mills goes out of her way to showcase products made in North Carolina with local ingredients. The wine list includes bottles from sixteen wineries; jams made by local cooks using Swain County berries; and locally made sauces, including ramp cornmeal from Smoky Mountain Heritage Products in Stecoah and Fire from the Mountain, cooked up by mountain farmers using their own hot peppers.

44 Frye Street, Bryson City (Swain County), 828-488-6207, www.thecottagecraftsman.com.

Brushy Mountain Bee Farm

While most of Brushy Mountain Bee Farm's beekeeping products are sold online, owners Steve and Sandy Forrest have set up a fascinating showroom for customers and the curious. The curvy mountain drive to reach it is part of the fun, too. Steve tends to twenty hives on the 120-acre property south of Wilkesboro and sells his honey at the shop. He also offers occasional beekeeping workshops. But for the most part, Brushy Mountain manufactures and sells beekeeping equipment. The showroom contains a little bit of everything, including "bee-ginners kits," hive frames, candle- and soap-making supplies, beekeeper clothing, and of course those cute plastic bear-shaped honey bottles. With business buzzing, the shop and packing and shipping rooms have expanded several times over, most recently in 2009.

610 Bethany Church Road, Moravian Falls (Wilkes County), 800-233-7929, www.brushymountainbeefarm.com.

DINING

The Admiral

When The Admiral opened in West Asheville in late 2007, it was a dive bar with great food. Now it's a hot spot for great food, with a dive bar. The Admiral has an ever-changing menu that follows the seasons. Its chefs go to the local farmers' markets (one is a few blocks away) a couple of times a week and cook up whatever is on hand; they also work with a few farms directly. "We got the reputation that we'd buy anything that people would bring us," said sous chef Drew Maykuth, "so farmers kept coming after

How They Treat the Animals We Eat

For "ethical omnivores," consumers who care about the welfare of the animals providing their food, several factors should be considered, including confinement, exposure to natural light and the outdoors, feeding, and slaughter methods. The three USDA-approved humane food certification programs are Certified Humane, American Humane Certified, and Animal Welfare Approved. All are backed by national animal advocacy organizations. Also, in 2009, Whole Foods Market launched its own animal welfare rating program.

The requirements of Certified Humane and American Humane Certified are similar, while Animal Welfare Approved policies are far more stringent. The latter is the only program to certify only family farms and to require that animals have access to the outdoors. The World Society for the Protection of Animals has an informative fact sheet on its website that includes charts comparing the three programs. Don't assume that if small farmers aren't certified, they don't treat their animals well, but do ask questions of them and ask to visit the farm to see for yourself how the animals are treated.

the markets with whatever they had left over. That's been really neat." The Admiral specializes in small plates, a little larger than tapas, which keeps the prices down on such dishes as chorizo sausage, mussels, and even duck—not what you'd expect to find inside a cinderblock tavern.

400 Haywood Road, West Asheville (Buncombe County), 828-252-2541, www.theadmiralnc.com. $–$$

The Blackbird

It turned out that Roz Taubman and Bobby Buggia were lucky they couldn't find the right place in Asheville to house a restaurant. Instead, in 2009, they opened The Blackbird in Black Mountain. It's close enough to woo Asheville food lovers but far enough away to quickly make its mark. "We wanted a relaxing, country destination–style restaurant," said Roz, a pastry chef who has worked on both coasts. The partners (he's executive chef) immediately became familiar with local products, including the heirloom grits from Peaceful Valley and the meat, cheese, and produce from the Foothill Family Farms collective and the Black Mountain Tailgate Market. The appreciation has gone both ways. "We've gotten tremendous community support from the beginning," Roz said.

10 East Market Street (Village of Cheshire), Black Mountain (Buncombe County), 828-669-5556, www.theblackbirdrestaurant.com. $$

Early Girl Eatery

Early Girl Eatery arrived early on the local-food scene in Asheville and nationally. Owners John and Julie Stehling opened the restaurant in 2001, years after meeting at the Hominy Grill in Charleston (owned by John's brother, Robert, who learned to cook at Crook's Corner in Chapel Hill). The couple immediately connected with area family farms, which still supply most of their produce and some cheese and meats. Early Girl, named after a hybrid tomato, focuses on southern dishes and also has become popular for its many vegetarian offerings. Thanks to its fresh food, moderate prices, and a steady stream of local and national publicity, Early Girl has rightfully kept its place in the spotlight for a decade.

8 Wall Street, Asheville (Buncombe County), 828-259-9292, www.earlygirleatery.com. $–$$

French Broad Chocolate Lounge

While cacao cannot be locally sourced, what a chocolatier puts inside a candy can be. At their French Broad Chocolate Lounge in Asheville, open since 2008, Jael and Dan Rattigan go to great lengths to work with area farmers. Their heavenly truffles are made with fruits, berries, honey, and eggs from area farms, along with herbs from their backyard garden. The milk they use is processed within a mile of their kitchen. The menu also includes cheese plates with selections from mountain cheesemakers such as Spinning Spider Creamery and Yellow Branch Farm.

10 South Lexington Avenue, Asheville (Buncombe County), 828-252-4181, www.frenchbroadchocolates.com. $

The Green Sage

Even for environmental- and locavore-minded Asheville, The Green Sage brought a gust of fresh air when it opened in 2008. The coffeehouse and café, started by Randy Talley and Al Kirchner, veterans in the organic grocery business, focus on sustainability all around. Almost all the food is local or organic or both, including beef and bison burgers and produce. To-go boxes are compostable, as is much of the dine-in waste, eliminating the need for a giant garbage bin out back. Perhaps the most prominent additions to the existing building, a former café, are the solar panels lining the roof, which supplies The Green Sage with hot water.

5 Broadway Street, Asheville (Buncombe County), 828-252-4450, www.thegreensage.net. $

Grove Park Inn Resort and Spa

When the Appalachian Sustainable Agriculture Project unveiled its now popular "Buy Appalachian" campaign in 2000 with a local-foods feast, it was no coincidence that the event was held at Grove Park Inn in Asheville. Even more so a decade later, the historic inn's Horizon and Blue Ridge dining rooms reach out to area farmers for their produce, poultry, and meat. One farmer in nearby Haywood County, where Blue Ridge chef de cuisine Denny Trantham hails from, plants five acres of vegetables exclusively for the restaurant's use. "We love the opportunity to showcase these great local products to all our visitors from near and far," he said.

290 Macon Avenue, Asheville (Buncombe County), 828-252-2711, 800-438-5800, www.groveparkinn.com. $$–$$$

Laurey's Catering and Gourmet to Go

Laurey's Catering and Gourmet to Go is the kind of place you'd want to support even if the food wasn't outstanding. Luckily, it is. Owner Laurey Masterton, a cancer survivor whose often touted motto is "Don't postpone joy," serves up fresh meals in her lively café in downtown Asheville. And, yes, she runs a full-service catering company. Masterton works with a long list of local farmers to incorporate local ingredients and products at every turn, an admirable feat for a high-volume business. She was a charter member of the county's Living Wage Campaign, and in 2009 she rode her bicycle cross-country to raise money for cancer research. We're happy to support her causes, especially when she's doing the cooking.

67 Biltmore Avenue, Asheville (Buncombe County), 828-252-1500, www.laureysyum.com. $$

Mamacita's

Rare is the inexpensive restaurant that uses any local products. Mamacita's in downtown Asheville is an exception, relying heavily on farm-fresh produce and even meat. Open since 2004, the popular spot serves what owner John Atwater has described as "Dixie-Mex." That label perfectly fits the "Veggie-Mama Burrito," filled with black beans, mashed sweet potato, and sautéed kale. The meat for the popular pork burrito comes from nearby Hickory Nut Gap Farm, and roma tomatoes and fresh tomatillos are roasted in house for the salsa. While Mamacita's prides itself on not being a chain, we wish a few more Mamacitas populated the state.

77-A Biltmore Avenue, Asheville (Buncombe County), 828-255-8080, www.mamacitasgrill.com. $

The Market Place

Since 1979 The Market Place has been not only one of Asheville's culinary centerpieces but the area's farm-to-table trailblazer. In 2009, owner-chef Mark Rosenstein decided it was time for a change. William Dissen, a few decades younger, was up for the challenge and bought the restaurant, keeping its name and identity. William is continuing the farm-fresh focus while adding modern twists of his own. "Fundamentally, our philosophies are the same, and we both used French-based classic techniques. I want to continue that while taking the restaurant toward the future." We hear a collective sigh of relief from Market Place fans near and far.

20 Wall Street, Asheville (Buncombe County), 828-252-4162, www.marketplace-restaurant.com. $–$$$

Table

Uptown and urbane, Table got its legs in 2005, when Jacob and Alicia Sessoms migrated to Asheville from New York to set up shop. Alicia, Jacob, and Matthew Dawes, who co-chefs with Jacob, all attended nearby Warren Wilson College. This new-American fine-dining spot has continued the farm-to-table approach it instituted when it first opened. Menu favorites include heirloom tomatoes with braised eggplant, sautéed local squash, and fregola; Hickory Nut Gap Farm pork ragu with zucchini; and orcchiette with tomato cream sauce.

48 College Street, Asheville (Buncombe County), 828-254-8980, www.tableasheville.com. $$–$$$

Tupelo Honey Café

Tupelo Honey Café opened in downtown Asheville in 2000 as a laid-back breakfast and lunch spot for southern comfort food with an emphasis on fresh, local ingredients. In its first decade, it grew into a tourist mainstay, adding dinner hours and a line of merchandise. In 2008 new owner Stephen Frabitore stepped things up even more, opening a second location and arranging a deal for a Tupelo Honey cookbook, to be published in 2011. Throughout this time, chef Brian Sonoskus has continued to draw customers with his creative, affordable dishes, many relying on area farmers. Much of the produce comes from Sonoskus's own Sunshot Organics, a twelve-acre farm he started in 2007. He grows vegetables, herbs, edible flowers, loads of blueberries, and even raises some laying hens. As for the Tupelo honey found on every table? That's from Florida, but we'll let it slide.

12 College Street, 828-255-4863; 1829 Hendersonville Road, 828-505-7676; Asheville (Buncombe County), www.tupelohoneycafe.com. $–$$

West End Bakery and Café

Even on a fall day at West End Bakery and Café, the blackboard that lists local ingredients was as packed as the lunch tables. Eggs, trout, basil, sausage, figs, chestnuts, paw paw, "some cucumbers," and more. "People in the neighborhood keep bringing in paw paws this week," explained co-owner Cathy Cleary. "I gathered chestnuts the other day at a neighbor's house." Baked goods are made with organic flour and local honey, and workers even planted a small garden behind the café with vegetables and herbs for cooking. From spring to fall, shopping is made easy, as the West Asheville Tailgate Market is held just next door.

757 Haywood Road, West Asheville (Buncombe County), 828-252-9378, www.westendbakery.com. $–$$

Zambra

The festive and vibrantly decorated Zambra opened with a splash in 2000 in downtown Asheville as a small Spanish tapas restaurant. Chef Adam Bannasch and manager Peter Slamp, who took over the restaurant a few years later, have expanded it in size, scope, and the use of local products, including produce, meat, cheese, and seafood. Though Zambra doesn't actively promote its local-farm sourcing, it is one of the most committed

Hopping for Hops

Western North Carolina has started to jump into hops production as a new specialty crop, something to supplement a farm's income. East Coast hops production is challenging because of humidity and disease, but farmers are greeting the challenge. Small hops operations have been started by more than fifty farmers in mountain counties. Now we're waiting for the beer-tasting tours.

restaurants in the city. "Adam is a really staunch believer in farm-to-table," Slamp said. "Like a tapas bar in Spain, we use the ingredients on hand. We're really a local tapas bar with a Spanish inspiration." With live flamenco and jazz in its lounge area, Zambra also is one of Asheville's most popular nightspots.

85 Walnut Street, Asheville (Buncombe County), 828-232-1060, www.zambratapas.com. $$

Square One Bistro

Joseph and Lindsay Lewis like a challenge. They got married, bought a house, and opened Square One Bistro all within a few months in 2008 and all before they'd turned thirty years old. Chef Joseph figures that 90 percent of his ingredients are local and organic during the growing season, and 60 percent otherwise, and he's devised a clever sourcing breakdown. Under his "50–150-regional" formula, produce comes from within fifty miles (some from his parents' nearby farm); "small protein," such as produce and poultry, is from within 150 miles; and "big protein," such as beef, is from the wider region. Square One has received glowing reviews not only for its sophisticated decor and creative dishes but also its optional small-plate offerings. In 2009 Square One opened a small takeout market that sells local produce, cheese, and meat, including its own homemade mozzarella and sausage.

111 South Main Street, Hendersonville (Henderson County), 828-698-5598, www.square1bistro.com. $$

Guadalupe Café

We nearly cried standing outside Guadalupe Café after discovering it was open for lunch only on Saturdays. We'd made a weekday noontime stop in adorable Sylva partly to eat at this funky little spot, in business since 2004 and anointed by *Gourmet Magazine* in 2007 as one of the country's top 100 farm-to-table spots. The restaurant and bar, frequented by Western Carolina University students, is owned by Jen Pearson, who immediately got to work establishing relationships with farmers throughout Jackson County. Her tamales, burritos, tacos, and more are filled with local produce, free-range meats, and sustainable seafood. Guadalupe also supports the local arts community by hosting live music and art exhibits. Next time, we'll check Guadalupe's schedule before making ours.

606 West Main Street, Sylva (Jackson County), 828-586-9877, www.guadalupecafe.com. $–$$

Knife & Fork

Unlike the many places that tout their use of local sources "when available," Knife & Fork in Spruce Pine turns that around by not using ingredients that aren't available locally. Well, except for lemons and celery, noted Nate Allen, who opened the restaurant with his wife, Wendy Gardner, in 2009. The economy brought them from their home in Los Angeles to the mountains of North Carolina, near Wendy's hometown of Burnsville. What drew them to tiny Spruce Pine was the legalization of the sale of alcohol within town limits—they needed to sell beer and wine to offset their higher food costs. Diners were lining up for tables within a week of the opening, eager to sample inventive dishes that raise the bar on farm-to-table dining.

61 Locust Street, Spruce Pine (Mitchell County), 828-765-1511, www.knifeandforknc.com. $$

Giardini Trattoria and Giardini Pasta and Catering Company

"The essence of Italian food is its freshness," said Mary Lyth, who, with her husband, Joe Laudisio, opened Giardini Trattoria on their small farm in the mountain town of Columbus. Here, freshness is only yards away in the couple's one-acre garden (*giardini* means "gardens" in Italian), visible from the restaurant's deck seating. It's planted with their essential ingredients—tomatoes, peppers, eggplant, garlic, and onion. Joe and Mary,

Nate Allen and Wendy Gardner rely on local farmers for almost everything they create at Knife & Fork in Spruce Pine (Mitchell County). Photo by Diane Daniel.

originally from Buffalo, first moved south to Hilton Head Island, South Carolina, where they bought an existing restaurant. When they moved to Polk County, they started Giardini Pasta and Catering Company in 2007, featuring gourmet takeout and wood-fired pizzas. In 2009 they opened a simple yet stylish twenty-seat dining room. A handful of area farms, along with their own, supply produce.

2411 Highway 108 East, Columbus (Polk County), 828-894-0234, www.giardinigardens.com. $$

The Purple Onion Café and Coffeehouse

Susan Casey learned to rely on the local harvest at an early age. Growing up on a tobacco and produce farm near Kinston, she would make lunch for the workers using ingredients from the farm. Today, as owner and chef of The Purple Onion Café and Coffeehouse in historic downtown Saluda, Susan still uses as many local offerings as she can get her hands on, including produce, fish, some meat, and of course apples in this orchard-filled region. "I do it because I just think it's so important." After it opened in 1998, not only did The Purple Onion become known for its Mediterranean and regionally inspired cuisine, but its live music on weekends has helped make it Saluda's most popular gathering spot.

16 Main Street, Saluda (Polk County), 828-749-1179, www.purpleonionsaluda.com. $$

M² Restaurant

When Myra Cowan opened M² Restaurant in downtown Spindale, next to Rutherfordton, she didn't advertise her local sourcing through Farmers Fresh Market, a local-source distribution service in Rutherford County. "I just thought it was something we should do. Sometimes you do something and assume everyone knows," she said. "Then we started to market it, and it's made a big difference in helping the business." The American bistro-style restaurant, open since 2007, is one of only a handful of restaurants in this small town, home to Isothermal Community College. "We have to offer a little bit of everything," Myra said, "but we do a lot of seasonal dishes." M²'s patio is especially lively for the restaurant's popular Sunday brunch.

125 West Main Street, Spindale (Rutherford County), 828-288-4641, www.msquaredrestaurant.com. $$

Gamekeeper

While Gamekeeper, which has been open since 1987, indeed serves game, including ostrich, venison, and buffalo, the Boone restaurant is now better known for its "modern mountain cuisine." Ken and Wendy Gordon bought the restaurant, housed in a cozy 1950s stone cottage, in 2000, and since then the enthusiastic couple has been paying homage to the region's farms and cuisine with an emphasis on organic and locally grown ingredients. High Country farmers praise the couple's support of local agriculture, which extends to using meat from humanely raised animals. For an extra treat, come to the tucked-away mountain setting in time to watch the sun set over the Blue Ridge Mountains.

3005 Shulls Mill Road, Boone (Watauga County), 828-963-7400, gamekeeper-nc.com. $$–$$$

LODGING

Zydeco Moon Farm and Cabins

A four-wheel drive will come in handy when you set out for Zydeco Moon Farm and Cabins, set on fifty acres north of West Jefferson. If you don't come equipped, don't worry, the farmers have two vehicles you can use to navigate the short but steep dirt road. "People get a kick out of that and always talk about it in the guest book," said Sally Thiel, who runs the six-acre certified organic produce farm with her husband, Joe Martin. They sell at farmers' markets and from the farm. The couple, who moved here from Baton Rouge, Louisiana, in 2005, give interested guests a farm tour. "Everybody gets a basket of tomatoes or whatever is growing," Sally said. The couple, who live in a farmhouse at the bottom of the hill, built the cabins, which come with vaulted ceilings and decks that look onto the Helton Creek Valley. And if you're itching to fish, Sally and Joe make that easy, too. Their stream is stocked with trout.

2220 Big Helton Road, Grassy Creek (Ashe County), 336-384-2546, www.zydecomoon.com. $$

Cloud 9 Farm

The slogan at Cloud 9 Farm—"Conservation, Celebration, Relaxation" is no exaggeration. Janet Peterson has transformed her 200-acre Fairview farm into an inviting retreat only twenty minutes from downtown Asheville. On the quiet grounds are two acres of U-pick blueberries, a small herd of hormone-free cattle, chickens, vegetable garden, educational apiary, and custom sawmill. Janet, whose family moved here in 1968, took over the farm after her parents died. In 2004 she turned the family home into a luxury vacation rental, and in 2007 she added a second cabin, equally well appointed, with furnishings made by local artisans. She also added a space for parties, weddings, and other special events. A former elementary school teacher, Janet was recognized by the North Carolina Forest Stewardship Program in 2008 for her commitment to natural resource conservation and management.

137 Bob Barnwell Road, Fletcher (Buncombe County), 828-628-1758, www.cloud9relaxation.com. $$–$$$

Hawk and Ivy Bed and Breakfast

James and Eve Davis have transformed a 1910 farmhouse on twenty-four acres into a lovely retreat about thirty minutes northeast of Asheville. Eve, a professional gardener and floral designer, is happy to lead tours through her large kitchen garden, featuring heirloom tomatoes, asparagus, peppers, and several potato varieties. She also grows herbs, fruit trees, and an assortment of berries. Much of the bounty will turn up on your breakfast plate. Since opening in 2004, Hawk and Ivy has become a popular wedding site, with ceremonies held in a huge field with mountain views or inside the reception hall, a former tobacco barn. James, an ordained minister, performs the ceremony, while Eve makes the bouquets.

133 North Fork Road, Barnardsville (Buncombe County), 828-626-3486, 888-395-7254, www.hawkandivy.com. $$

Ox-Ford Farm Bed and Breakfast Inn

"That house called me," said Edith Hapke, pointing to the farmhouse she saved from ruin. Dr. Hapke, as she's known in these parts from her decades of work in Asheville as a pulmonologist, bought the 1860s farmhouse in 1972, having come to the United States from Germany as a college student. Twenty years later she opened the Ox-Ford Farm Bed and Breakfast Inn. Dr. Hapke also is a working farmer, raising sheep, cattle, and chickens on steep mountain pastureland. Her heavy lifting is even more impressive given her age, which she gives as "ancient." Guests may involve themselves in farm activities or simply sit on the beautifully landscaped lawn or front porch and relax to the sounds of the nearby creek.

75 Ox Creek Road, Weaverville (Buncombe County), 828-658-2500, www.ox-fordfarm.com. $$

Bedford Falls Alpaca Farm

Doug and Nancy Rondeau moved from Florida to the far southwestern mountains of North Carolina in 1999 to raise alpacas, making their Bedford Falls Alpaca Farm the first in the state, said Nancy, who continues

Boyd Mountain Christmas Tree Farm in Waynesville (Haywood County) offers choose-and-cut trees as well as lodging in historic cabins. Photo by Selina Kok.

to show and breed the now popular animals. "I just love them," she said. "They're sweet and gentle creatures. The books say they're intelligent. Sometimes they are." Since 2004, guests have been able to stay in a spacious and modern cabin, with a back deck overlooking the alpaca pastures. On the other side of the cabin, a deck overlooks a flowing creek. "Guests can watch them, and I'll take them to the barn to feed them," Nancy said. For a free weeklong stay, all you have to do is buy an alpaca.

171 Pine Trail, Warne (Clay County), 828-389-1345, www.bfaf.com. $$

Boyd Mountain Christmas Tree Farm

You don't have to wait for the holidays to surround yourself with Christmas trees. Boyd Mountain Christmas Tree Farm, a choose-and-cut and wholesale Fraser fir farm in Waynesville, also offers outstanding lodging. Dan and Betsy Boyd, who started growing trees in 1984, have seven cabins on their 130 acres, all retrofits of hand-hewn Appalachian log cabins dating back 150 to 200 years. Dan, a retired dentist, had long been fascinated by log cabins, rescuing logs from decaying homesteads in western North

Carolina and eastern Tennessee. On the farm, he reassembled the cabins and added whatever modern structural assistance was needed. They feature all the modern amenities and have lovely views, some of mountains, others of landscaped lawns, and most including rows of Christmas trees.

143 Boyd Farm Road, Waynesville (Haywood County), 828-926-1575, www.boydmountain.com. $$–$$$. Choose-and-cut open late November to December.

The Gardener's House at Frog Holler

Frog Holler Organiks outside of Waynesville is an organic produce farm mostly serving the Cammer family, but they're happy to share some of the bounty with their guests at The Gardener's House. The spring-to-fall rental (and wedding space) is in an attractive 1920s farmhouse outfitted as a modern vacation home. The view is of pastures and mountains, but Interstate 40 is very close, so the sound-sensitive should be aware. Along with its large garden, the farm is home to cows, horses, and chickens, which guests are welcome to visit up close. While guests cook their own breakfast, the farm chickens supply the eggs.

234 Tommy Boyd Road, Waynesville (Haywood County), 828-627-3363, 828-400-0419, www.frogholler.net. $$

Bend of Ivy Lodge

When Walker and Doug Silsbee converted a fifty-year-old tobacco barn into Bend of Ivy Lodge, a retreat center, they preserved the structure and its history. The original support beams remain in use, and thick poles that once hung in the rafters for drying tobacco now serve as artful balcony railings on the second floor. Some walls are decorated with antique farm tools "to remind people they're in a barn," Walker said. "We really wanted to honor the place, and its past." The lodge, on sixty-three acres outside of Marshall, sleeps twenty-six people and can be rented for group retreats, weddings, and, occasionally, personal retreats.

3717 Bend of Ivy Road, Marshall (Madison County), 828-645-0505, 888-658-0505, www.bendofivylodge.com. $–$$$

Briar Rose Farm

Tom Hare and Judie Hansen moved to the North Carolina mountains northwest of Asheville in 1994 to escape Chicago's winters. Things really took off when Judie, a former visiting nurse, asked Tom, a jack-of-all-trades, "What kind of animals can I buy?" He answered, "You can buy anything you want." Years later he told us, with a wide grin, "That was a mistake." Tom's weak moment is now the delight of many a family who stay at one of three cabins here, where they can get up close and personal with beautiful Belted Galloway cows, goats, and laying hens. From the base of an old farmhouse, Tom built one large rental cabin near the couple's home and then added two others (attractive manufactured varieties) in scenic settings on nearby hills on the farm's 260 acres. U-pick blackberries and vegetables are available for the "cabin people," as they call their guests. We call them lucky.

91 Duckett Top Tower Road, Hot Springs (Madison County),
828-622-7329, www.briarrosefarm.com. $$

Broadwing Farm Cabins

Mary and Pete Dixon have been growing vegetables, fruits, herbs, and plants in Hot Springs since 1992. Their latest offerings have been greenhouse-grown heirloom tomatoes, perennial blackberries, apples, and potted herbs. You can see their sustainable practices for yourself and purchase whatever's in season if you stay in one of the three rental cabins on the eighty-five-acre farm. While the Dixons don't lead farm tours, renters are welcome to wander the grounds. Each cabin has a deck, two with views of the French Broad River valley and the mountains beyond and one set back a bit in the woods. And of course they don't call this place Hot Springs for nothing. Each cabin has a hot tub—filled with naturally warm spring water.

Berry Wills Road, Hot Springs (Madison County), 828-622-3647,
www.broadwingfarmcabins.com. $$

The Cottages at Spring House Farm

The Cottages at Spring House Farm, in the foothills of the Appalachians south of Marion, is a ninety-two-acre sanctuary where the present connects with the past. Thanks to the restoration work of owners Arthur and Zee Campbell, the farm setting has been left intact, and, in some cases, remarkably restored. The couple bought the land in the late 1990s and opened for business in 2000. Visitors first arrive at the Campbells' home, a circa 1826 farmhouse now on the National Register of Historic Places that they're happy to show guests. Five contemporary cabins were built with salvaged wood from this and surrounding farms. But our favorite is the Appalachian, a circa 1835, two-story cabin that was moved from land nearby. Most of the historic features remain, including fifteen-pane windows and wide-planked heart-pine floors and walls. While the other cabins are wooded, this one has a large lawn abutting a pasture. The August day we visited, rolled bales of hay dotted the field, underscoring the country feel.

219 Haynes Road, Marion (McDowell County), 877-738-9798,
www.springhousefarm.com. $$–$$$

Laurel Oaks Farm

Al Bessin used to work for Dole Food Company in human resources, and he and his wife, Yvonne, endured seventeen corporate moves. So they were happy to settle down in the mountains north of Bakersville in 1991. On sixteen acres of pasture on their mostly wooded 200-acre Laurel Oaks Farm, the couple tends to a couple of dozen Tunis, a historic breed of sheep. The Bessins' comfortable rental property abuts the pasture, and sheep are often in sight. "We do a big thing for guests at five o'clock when we bring the sheep in," Yvonne said. "The kids can pet and hand-feed them." The animals aren't just for show. A typical shearing produces 200 pounds of fleece, some of which Yvonne spins, dyes, and knits. She's happy to give spinning demonstrations, and she sells raw wool fleece, yarn, and beautifully knit products.

7334 Highway 80, Bakersville (Mitchell County), 828-688-2652. $$

Green River Vineyard Bed and Breakfast

All we could say when we pulled up to Green River Vineyard Bed and Breakfast was, "Wow." Owners Peggy and Claude Turner have outdone themselves. In 2003 Peggy, a registered nurse, and Claude, a retired air traffic controller, moved here from Charlotte, building their home and clearing five acres on a thirty-acre wooded lot between Rutherfordton and Columbus. The B&B opened the following year. The wide mountain views are stupendous, without a house in sight. Claude is the grape guy and is a sought-after mentor and grower, selling to several foothills wineries. Peggy, who also helps in the vineyards and nurses part-time, runs the beautifully appointed bed and breakfast. She tends to a large garden, which she harvests for breakfast items and afternoon hors d'oeuvres.

3043 John Watson Road, Green Creek (Polk County),
828-863-4705, www.greenriverbb.com. $$

Taking Lawn Maintenance by the Horns

Farmers have long used livestock to clear pastures and brush, but in the past few years the natural mowing method has spread to the city.

In Durham, Alix Borman in 2008 combined her business and naturalist backgrounds to unleash Goat Patrol (www.thegoatpatrol .com), a small herd of goats she takes to the yards of Triangle residents during the day to chow down on such delicacies as poison ivy, pines, cedars, English ivy, honeysuckle, wisteria, and kudzu. It didn't take long for neighbors to organize goat-watching gatherings, turning the Goat Patrol into a goat party.

At Wells Farm (www.wellsfarmgoats.com) near Brevard, in the mountains, longtime farmers Ron and Cheryl Searcy first bought a few goats to clear their own land more than a decade ago. Several years later, after taking the goats to help a neighbor clear brush, they decided to start a brush-clearing business. Grazing requests grew, as did the herd, to more than 200 goats, and the Searcys' tribe serves homeowners, businesses, municipalities, states, and conservation groups. Wells Farm goats usually punch the clock in North Carolina and nearby, but they've commuted all the way to New York. That's upstate, not Manhattan. At least not yet.

The Mast Farm Inn

For more than a century, visitors to the mountains have stayed at The Mast Farm Inn near Boone. In 2006 several members of an extended French family took over operations, continuing the inn's relationship with the land and history of rural Valle Crucis. Seven guestrooms are available in the original 1880s farmhouse, which is graced with an inviting wraparound porch. Eight other cabins and cottages on the grounds are rented out as well, some new, others historic, including the Loom House, the two-room log cabin that started the farm in the early 1800s and was later turned into a space for spinning and weaving. Guests are invited to stroll through the inn's organic garden, which supplies its outstanding farm-to-table restaurant, Simplicity. A dinner here is worth the trip alone. And make sure you stop at Simplicity's Pantry, a small shop filled to the brim with food products made in North Carolina and Virginia.

2543 Broadstone Road, Banner Elk (Watauga County), 828-963-5857, 888-963-5857, www.mastfarminn.com. $$–$$$$

The Old Farmhouse

Don't let the name fool you. While this is a century-old farmhouse, it has been beautifully restored and comes with all the modern amenities. Hardwood floors, wooden cabinetry, handmade quilts, and vintage photographs make this a real rural retreat. Owner Walton Conway and his family, who run Golden Cockerel in Boone, an exporter of Russian nesting dolls, live across the pasture in a more modern home. The sixteen-acre spread is set in a valley at the base of Bald Mountain and near Elk Knob State Park. Chickens, tail-wagging dogs, a donkey, and a pot-bellied pig add to the scenery. Guests can help gather eggs and, by appointment, ride the farm horses. Nonguests may schedule a tour, fresh eggs included.

1346 Willet Miller Road, Todd (Watauga County), 828-297-4653, 828-297-9571, www.goldencockerel.com. $$

Songbird Cabin

Driving east of Boone along the Watauga River to the breathtaking Mountain Dale community has been a seasonal ritual for regulars coming to Cornett Deal Christmas Tree Farm since the late 1980s. In 2000 owner Diane Cornett Deal gave folks a reason to visit year-round when she built an attractive vacation cabin atop a hill on her twenty-six-acre farm. The many windows look out onto views of mountains and rows of Fraser firs. A Christmastime holiday shop features locally made knitted goods and other items, including Diane's own hand-thrown pottery and a selection from the hundreds of wreaths she and her parents make every year.

142 Tannenbaum Lane, Vilas (Watauga County), 828-964-6322, www.songbirdcabin.com. $$. Choose-and-cut open late November to December.

SPECIAL EVENTS AND ACTIVITIES

Log Cabin Cooking and Music

Old-time cooking expert Barbara Swell spent much of her twenties living on farms. "After college, the thing to do was farmhouse-squatting off the grid," she said. Barbara is now living on the grid in Asheville, where she and her husband, Wayne Erbsen, have turned their passion for southern Appalachian cooking, music, and folklore into a cottage industry. Wayne performs and teaches old-time music, while Barbara has written many books about historic cooking and teaches cooking out of the log cabin next to their home. From her kitchen that looks like grandma's, Swell teaches traditional cooking skills, including making pies, jams, jellies, and butters, as well as hearth cooking, often using ingredients from her extensive home garden.

111 Bell Road, Asheville (Buncombe County), 828-299-7031, www.nativeground.com.

Mountain State Fair

The capital city isn't the only place to hold a state fair. Since 1994, the Mountain State Fair, sponsored by the North Carolina Department of Agriculture, has brought the midway to western residents. The ten-day event, offering much of what Raleigh has but on a smaller scale, has the more down-home feel of a county fair. Agriculture is underscored, with more display animals and livestock events, including one of the largest dairy goat shows on the East Coast. The music is more country, too, with a bluegrass and old-time music competition. In 2009 the fair added new attractions, including a chairlift ride that hovers over the crowds and new million-dollar buildings. One of the latter, the Virginia C. Boone Mountain Heritage Center, is a 12,000-square-foot log-frame building housing Appalachian artisans. The opening-day celebrity was nonagenarian Virginia Boone herself, braiding rugs as usual.

1301 Fanning Bridge Road (WNC Agricultural Center), Fletcher (Buncombe County), 828-687-1414, www.mountainfair.org. Held in September.

True Nature Country Fair

Inspired by the famed Common Ground Country Fair in Maine, the True Nature Country Fair, organized in 2007 by the folks at the nonprofit Organic Growers School, celebrates down-home goodness with a sustainable and activist bent. At the annual event at the Big Ivy Community Center in Barnardsville, southern Appalachian farmers, homebuilders, craftspeople, restaurateurs, and social activists join to spread their gospel, and wares, to their neighbors. Everything featured at the fair is from the southern Appalachian region and is produced from resources that are at least 50 percent organic or sustainable. Exhibit areas typically include a farmers' market, craft displays, and information on energy, shelter, and wellness, with workshops on gardening, farming, and homesteading.

540 Dillingham Road, Barnardsville (Buncombe County), www.organicgrowersschool.org. Held in September.

John C. Campbell Folk School

From fruit-tree grafting to cooking with fresh produce, the nonprofit John C. Campbell Folk School in the state's far western mountains has a terrific selection of agriculture-related courses and workshops. "Winter Harvest Kitchen" teaches seasonal growing and cooking, "Appalachian Shawl: From the Pasture to the Pattern" takes students from sheep to shawl, and food preservers will appreciate "Canning Chutneys, Cordials, and Condiments." The school, which has celebrated Appalachian farmers and artisans since 1925, also tends to large vegetable and herb gardens, used for classes and the student dining hall.

1 Folk School Road, Brasstown (Clay County), 828-837-2775, 800-365-5724, www.folkschool.org.

Annual Ramp Convention

Ramps festivals have become popular throughout the eastern mountain states, but few if any towns have been celebrating the pungent Appalachian wild leeks longer than Waynesville. The Annual Ramp Convention, sponsored by the American Legion Post 47, started in the 1930s. Every May it draws thousands to sample the ramps, cooked in dishes and eaten raw or pickled. The down-home festival also features country and bluegrass bands, local cloggers, and a bake sale. In the past few years, some area farmers have started to cultivate ramps, which are in the same family as onions, garlic, and scallions. Over in Stecoah, near Robbinsville, the Smoky Mountain Native Plants Association makes and sells ramp cornmeal.

171 Legion Drive, Waynesville (Haywood County), 828-456-8691. Held in May. Ramp cornmeal information, 828-479-8788, www.smnpa.org.

North Carolina Apple Festival

A is for the more than a dozen varieties of apples on hand during the annual North Carolina Apple Festival, held every Labor Day weekend in downtown Hendersonville. Started in 1946, the event honors growers in Henderson County, the largest apple-producing county in the state. Visi-

Help for Farmworkers

Many farms hire helpers during the highest months of production. Sometimes these are interns and area residents, and other times they are migrant workers who travel from place to place. North Carolina ranks sixth in the nation in the number of migrant workers it employs, with an estimated number of at least 150,000 farmworkers and dependents here each growing season, according to the North Carolina Farmworker Institute (www.ncfarmworkers.org). The institute is a project of the Farmworker Ministry Committee of the North Carolina Council of Churches and includes several advocacy groups statewide.

Most migrant workers are immigrants from Spanish-speaking countries and are here under the federal H2A "foreign guest worker" visa program. Typically their work is hard and their pay is low, with few wage protections and benefits. Other issues social justice groups are working to improve include substandard and overcrowded housing, pesticide exposure, high rates of illness, limited workers' compensation, and limited access to health care.

Over the years, some improvements have been made. In 2004, a historic labor agreement spurred by a five-year boycott of Mt. Olive Pickle Company was signed by the Farm Labor Organizing Committee, North Carolina Growers Association, and Mt. Olive Pickle, unionizing H2A guest workers for the first time in the country. And in 2007, the Migrant Housing Act of North Carolina marginally improved migrant living conditions, but there is still much work to be done.

tors also can see the trees up close during an orchard tour. Other festivities include a nine-block street fair with more than 150 craft and food vendors, continuous live music and, on Labor Day, the King Apple Parade. All combined, it's enough apples to keep the doctor away until the next year's festival.

Downtown Hendersonville (Henderson County), www.ncapplefestival.org, 828-697-4557. Held Labor Day weekend.

Organic Growers School

The Organic Growers School isn't a year-round school but a springtime weekend event at various sites in western North Carolina. The grassroots organization grew out of the volunteer efforts of a group of farmers and extension specialists who gathered in 1993 to discuss the need to supply affordable crop-growing information to western North Carolina farmers, as well as to impart information on organic and sustainable agriculture. Its first spring conference, a day of workshops, attracted about 100 participants. Now it draws more than 1,300 students, who choose from more than seventy classes on such topics as starting a vegetable garden and raising goats. In 2007 the nonprofit organization started the True Nature Country Fair, and in 2009 it created two new programs—the Farmer Training Initiative, to pair established farmers with students of farming; and the N.C. Apprentice Link, a database that connects farms seeking labor with willing interns.

Locations vary. www.organicgrowersschool.org. Held in March.

Madison Family Farms

Madison County farmer Dewain Mackey thought the county could be doing more to help farmers after the tobacco buyout, so he hatched a plan. In 2006 he created Madison Family Farms, a nonprofit organization that connects local growers to consumers and establishes ties between farmers and the community. Mackey ran the operation, under the direction of the extension service, until 2009, when it hired its first full-time director. Its most popular public program is Harvest Meals, a fall series of catered meals sourced by Madison farms. The organization also runs a farm-to-school program and maintains a food-processing center where farmers can make value-added products, such as jams and sausages. In 2009 a shiitake mushroom cooperative was formed and a shed was built to house the logs. If you're looking to sample some of the farmers' products, check out Madison Family Farms' gift baskets, which range from edible items to lotions, soaps, and bath salts.

258 Carolina Lane, Marshall (Madison County), 828-649-2411, www.madisonfarms.org. Tours by appointment.

Family Farm Tour

The most scenic farm tour in the state has got to be the Family Farm Tour in western North Carolina. Organized by the Appalachian Sustainable Agriculture Project, the self-guided weekend tour includes farms in Buncombe, Yancey, Henderson, Madison, Haywood, and Transylvania counties. Farms are clustered in six geographical groups for ease of visiting, as this tour covers significant distances, especially if you factor in mountain roads. This tour also might have the most diverse selection among its nearly forty stops, including bison, trout, fiber, lavender, Christmas tree, and goat cheese farms.

828-236-1282, www.familyfarmtour.org. Held in June.

High Country Farm Tour

The High Country Farm Tour typically features more than a dozen farms stretching over five mountain counties—Watauga, Ashe, Alleghany, and Wilkes in North Carolina and Grayson just over the state line in Virginia. So while you can't visit them all unless you're up for a lot of driving, the scenery is sure to be splendid in whichever region you choose. The tour is organized by Blue Ridge Women in Agriculture, a Boone-based nonprofit organization of farmers, gardeners, businesses, and individuals committed to creating a food and farming system that celebrates the producers and consumers of local food.

www.brwia.org. Held in August.

RECIPES

Rustic Creamy Apple Pie

Lindsey Butler of Sky Top Orchard in Henderson County calls this simple pie recipe "the ultimate comfort food that takes only minutes to make. Start it an hour or so before dinner, and it will be ready for dessert when you are."

SERVES 8 TO 10

1	unbaked 9-inch pie crust (purchased or homemade)
1	cup whipping cream (unwhipped)
2	tablespoons cornstarch
4 to 5	full-flavored (about 6 cups) sweet/tart apples, sliced ⅛ inch thick
¾	cup sugar
½	teaspoon ground cinnamon
½	tablespoon butter, cut up

Preheat oven to 375 degrees.

Lightly butter a 9-inch glass pie pan. Roll out the dough and place in the prepared pie pan. Decoratively crimp the edges. Refrigerate until ready to use.

In a small bowl combine ⅓ cup of the whipping cream and cornstarch. Mix well until no lumps remain.

Place the remaining whipping cream in a small saucepan and place over medium heat. Add the whipping cream/cornstarch mixture and continue to cook until the cream has thickened. Remove from the heat and cool to room temperature.

In a large bowl combine the apples, cream, sugar, and cinnamon and stir to coat the apples. Pour the apples into the prepared pie pan. Dot with the butter. Bake in the bottom third of the oven for 1 hour and 15 minutes or until the pie has puffed up, the apple slices are tender, and the crust is golden brown around edges. Check the pie after 1 hour and cover with buttered foil if it is browning too quickly.

Remove the pie from the oven and let cool for 40 minutes before serving.

Oven-Fried Chicken

This moist and crunchy chicken dish is a perfect example of the "gourmet comfort food" Laurey Masterton serves at Laurey's Catering and Gourmet to Go in downtown Asheville. Laurey often uses chicken from nearby Hickory Nut Gap Farm.

SERVES 6

3	cups cornflakes, crushed to a rough crumble
1 ½	cups all-purpose flour
1 ½	teaspoon kosher salt
1 ½	teaspoon freshly ground black pepper
	Pinch of cayenne pepper
2	tablespoons chopped fresh thyme or marjoram
1	cup well-shaken buttermilk
2	eggs
	Dash of hot sauce
6	boneless, skinless chicken breasts
6	ounces peanut oil

Preheat oven to 325 degrees.

In a shallow dish combine the cornflakes, flour, salt, pepper, cayenne, and herb. In another shallow dish combine the buttermilk, eggs, and hot sauce, and stir to combine.

Dip the chicken breasts in the buttermilk mixture and then the cornflake mixture, thoroughly coating each piece of chicken.

Heat the peanut oil in a large frying pan or cast-iron skillet over medium-high heat. Add the chicken in batches and fry until the breasts are golden brown, about 2 minutes per side. Remove from the pan and drain on paper towels. Place the chicken on a large baking sheet and put in the preheated oven to bake for 20 to 25 minutes, until chicken is cooked through. Serve hot or at room temperature.

Sunburst Grilled Apple Ginger Trout Fillets

Sally Eason of Sunburst Trout Company in Haywood County says this simple trout recipe is her long-standing favorite. If your local grocer doesn't carry Sunburst Trout, you can order it online.

SERVES 4 TO 6

1	large can frozen apple juice concentrate, thawed
2	cloves garlic, minced
2	tablespoons grated fresh ginger
¼	cup balsamic vinegar
¼	cup light soy sauce
1	small bunch green onions, chopped (using the whole onion)
4 to 6	8-ounce Sunburst rainbow trout filets
	Salt and freshly ground black pepper

Place the first 6 ingredients in a large bowl and stir to combine. Place the filets in a large baking dish and pour the marinade over the top. Cover in plastic wrap and refrigerate for at least 1 hour or up to overnight. Marinating overnight will intensify the flavors.

Light a grill. Season the trout with salt and pepper. Place filets skin side down and grill for about 4 minutes. Turn and cook for another 2 minutes or until lightly charred and cooked through.

Serve immediately.

Heirloom Tomato Cobbler

Chef Sara Hord's tomato cobbler is a special way to celebrate the summer's bounty of heirloom varieties at her Millstone Meadows Farm in Morganton and is a favorite at her farm dinners.

SERVES 6 TO 8

CRUST

1 ½	cups all-purpose flour
1 ½	teaspoons sea salt
½	cup cornmeal, white or yellow
2	teaspoons baking powder
¼	cup grated parmesan cheese, about 1 ounce
1	stick cold butter, cut into pieces
3	tablespoons lard or nonhydrogenated vegetable lard
⅔	cup buttermilk

Place the first 5 ingredients in a large bowl and stir to combine. Using a pastry blender or your hands, work the butter and lard into the flour mixture until it reaches the consistency of small peas. Gradually add the buttermilk and bring the dough together with a fork until it forms large clumps. Turn the dough onto a lightly floured work surface and knead lightly about 4 to 5 times. Do not overwork the dough. Pat the dough into a flat disk, wrap in plastic wrap, and chill until ready to use.

TOMATO FILLING

1	tablespoon olive oil, plus more for oiling the baking dish
2	medium sweet onions, halved and thinly sliced
3 to 4	pounds medium heirloom garden tomatoes, sliced ¼-inch thick
⅓	cup quality mayonnaise
2	tablespoons cornstarch
2	teaspoons sea salt
	Freshly ground black pepper
2	tablespoons fresh basil, chopped
½	cup fontina cheese, grated (about 2 ounces)
½	cup white cheddar cheese, grated (about 2 ounces)
1	egg, lightly beaten
1	tablespoon milk
1	teaspoon fresh Italian parsley, chopped

Preheat oven to 350 degrees.

In a large skillet, heat the olive oil over medium heat. Add the onions and sauté until soft, about 5 to 7 minutes. Place the onions and tomatoes in a bowl and toss with mayonnaise, cornstarch, salt, pepper, and basil. Oil the bottom and sides of a 9 × 13-inch baking dish. Layer half of the tomato mixture on the bottom of the baking dish, then sprinkle with half of the cheeses. Repeat with the remaining ingredients.

On a lightly floured surface, roll the dough to ½-inch thick. Cut the dough into pieces with a dough cutter and place on top, leaving open spaces for the tomato mixture to bubble through. In a small bowl, make a glaze with the beaten egg and milk. Brush the pastry with the glaze and sprinkle with parsley and pepper.

Bake for about 45 minutes. Check the cobbler after 30 minutes and cover with aluminum foil if it is browning too quickly. Cool and serve.

Sesame Kale

Tryon cookbook writer Keith Snow has the solution for you CSA subscribers who aren't sure what to do with the large amounts of kale in your weekly boxes of produce. This Asian-inspired side dish from his Harvest Eating Cookbook dresses up the healthy vegetable and might even have you asking for more.

SERVES 4

2	pounds kale (3 cups steamed)
2	tablespoons light or pure olive oil
1	medium shallot, sliced
3	tablespoons low-sodium soy sauce
2	teaspoons toasted sesame oil
2	tablespoons black and white sesame seeds

Steam the kale in a stove-top steamer for about 10 minutes. Run the kale under cold water to stop the cooking process. Let it drain and set aside.

Heat the olive oil in a wok or sauté pan over medium heat. Add the shallot and sauté for 1 minute. Add the steamed kale and stir-fry for 1 minute. Add the soy sauce, sesame oil, and sesame seeds and stir-fry for 1 minute longer. Serve warm.

From *The Harvest Eating Cookbook: More Than 200 Recipes for Cooking with Seasonal Local Ingredients*, by Keith Snow. Reprinted by arrangement with Running Press, a member of the Perseus Books Group. Copyright © 2009.

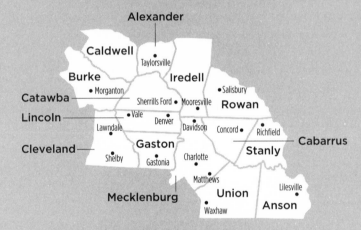

Alexander

Caldwell

Burke

Catawba

Lincoln

Cleveland

Taylorsville

Iredell

● Morganton

Sherrills Ford ● Mooresville

● Salisbury

Rowan

● Vale

Denver

Davidson

Concord ●

● Richfield

Cabarrus

Lawndale

Gaston

Shelby

Gastonia

● Charlotte

Stanly

Mecklenburg

Matthews

Union

Waxhaw

Lilesville

Anson

Charlotte Area

The Charlotte area grew and prospered from the harvests of eastern cotton farmers. By the 1900s, the Queen City was known for its textile manufacturing plants. As more mills relocated to outlying areas, Charlotte's growth came from finance and distribution. Now, as a mostly urban and suburban fourteen-county area, its farms are few but notable for their diversity, raising everything from pumpkins, to heritage livestock, to ostriches, to organic produce. Mecklenburg County has an enviable supply of farmers' markets—the state-run Charlotte Regional Farmers Market and the community-minded small-town markets in Matthews and Davidson. A growing number of chefs play their part as well to please local-food fans.

FARMS

Creekside Farms

Chad and Faith VonCannon had personal reasons for starting their forty-acre Creekside Farms, near Mount Pleasant. "We wanted our son to be raised on a farm, and we wanted people to see where their food comes from," said Chad, a civil engineer. The young couple, who began dating while at Mount Pleasant High School and then continued throughout their time at North Carolina State University, have farming backgrounds. They bought their land in 2007 and started selling meat at farmers' markets in 2009. "We got some cows, then goats, and like more and more people, we wanted to grow as much of our own food as possible," Chad said. They typically have about eighty laying hens, a few hundred broilers, pigs, and goats. They also raise grass-fed beef cows on a family-owned pasture nearby. The way Chad sees it, "a whole generation missed out on farming. Now people in their thirties are bringing their children to farms to see what they didn't get to see." The couple plans to open a farm store but until then will sell from the farm by appointment.

7611 Mount Olive Road, Concord (Cabarrus County), 704-436-5320, www.creeksidefarms-nc.com. Sales and tours by appointment.

Elma C. Lomax Incubator Farm Park

Some thirty miles northwest of Charlotte, a forward-thinking farm project has been incubating since 2009. The Elma C. Lomax Incubator Farm Park in Cabarrus County is a thirty-one-acre spread parceled into half-acre plots for new farmers to use. The property was bequeathed to the county in 2002 to be used as a park. Once the farm is established, trails, picnic shelters, and an educational center will be added. For now, it's a place for farmers to learn how to use equipment, work the land, and sell their harvest. Several freshman farmers already have started a CSA from there, among them Aaron Newton, coauthor of *A Nation of Farmers*. The county and the Cabarrus cooperative extension service oversee the farm. Occasional events are held here, with more expected as this progressive park progresses.

4335 Atando Road, Concord (Cabarrus County), 704-920-3310.

Maple Lane Homestead

"Hey sheep! Come on, girls!" yelled Kelly Foster, hoping some of her two dozen Icelandic sheep would appear. Distant bleating signaled their location, and they slowly came into view. Kelly and her husband, Doug, also raise dairy and cashmere goats and heritage breeds of chickens on their beautiful thirty-acre Maple Lane Homestead in Concord, which they started in 1992. "I grew up on a farm," Kelly said. "The green beans had to be picked before my sister and I could go out on dates on Friday night." Depending on the season, she might be selling meat, eggs, produce, or fiber. Kelly also is known for the organic bread she bakes for The Bradford Store, which often sells out while the loaves are still warm.

8752 Overcash Drive, Concord (Cabarrus County), 704-782-6627, www.maplelanehomestead.com. Sales and tours by appointment.

Riverbend Farm

A trek to Riverbend Farm, forty minutes east of Charlotte, is an autumnal ritual for many families in the area. Since 1984 owners Jim and Mary Little have continued to add to the offerings on their ninety-two-acre farm, which include homegrown pumpkins, buffalo, goats, a playground, gemstone panning, an ice cream shop, and a whopping twenty-two-acre pumpkin patch. And don't forget the giant cow. In 2009 the couple spied a ten-foot fiberglass Holstein in the Midwest and bolted it to a trailer for the ride home. "You should have seen the looks we got on the road," said Mary. "My husband is nothing but a big kid." Other highlights are a log barn Jim reassembled from another site and a "country store" featuring vintage heart-pine flooring made from church pews. Rocking chairs sit on the wide front porch, inviting parents to rest while the kids wear themselves out.

12150 McManus Road, Midland (Cabarrus County), 704-888-2891,
www.riverbendfarm.net. Open late September to October.
Country store available for parties.

Twelve Acre Academy

Even before she had her two children, Loretta Loughlin would help other mothers with home schooling by chaperoning field trips. Since 2006 she's led her own tours at Twelve Acre Academy, which she established on the small farm she and her husband own in Mount Pleasant. Alpacas and goats are center stage. Children can feed and walk the alpacas and milk a goat. Loretta shears the alpacas herself and spins her own yarn. "We have shearing days in the spring and I try to have an alpaca for each group to see."

401 Dutch Road, Mount Pleasant (Cabarrus County), 704-436-8855,
www.twelveacreacademy.net. Sales and tours by appointment.

BirdBrain Ostrich Ranch

"That would make a heck of an omelet, wouldn't it?" said Pat Roberts with a grin as she passed us an egg weighing more than two pounds. Since 1993 she and partner Mike Todd have tended to an ever growing flock of ostriches at their BirdBrain Ostrich Ranch, an hour north of Charlotte. Watching the largest birds in the world — up to eight feet tall and weighing around 250 pounds — strutting around a fenced-in pasture off a Carolina country road is a sight to see. The couple keep upward of fifty birds, from chicks to adults, and take six to eight to market once a month or so. They not only sell the meat (which is red and lean) but decorate egg shells, make dog bones, and tan some of the hides. Pat also collects historical ostrich memorabilia and ostrich science information. "I'm convinced that they're dinosaurs," she said.

6691 Little Mountain Road, Sherrills Ford (Catawba County), 704-483-1620, www.birdbrainranch.com. Sales and tours weekends or by appointment.

HarvestWorks

At HarvestWorks near Shelby, children and adults with disabilities can work on a small farm set on thirty-two acres. They tend to a produce garden, a handful of animals, and even hundreds of tilapia and blue gill in the fish farm. The public is invited to walk around the farm, the perfect scale for young children. Animals include miniature horses, donkeys, goats, and a full-grown turtle named Franklin. Tractor tours take visitors past the gardens, greenhouses, and through the surrounding wooded trails, and a kid-friendly cartoon-oriented Halloween trail is held every October. A small country store sells crafts made by HarvestWorks clients and sometimes produce from the farm.

891 North Post Road, Shelby (Cleveland County), 704-487-7777, www.harvestworksinc.org. Open daily. Tractor tours by appointment.

Apple Orchard Farm

After retiring from a career in business management at Duke Energy in 2004, Art Duckworth plunged headfirst into sustainable farming on land that has been in his family for decades. On sixty-five acres northeast of Charlotte, which Art calls Apple Orchard Farm, he tends to heirloom varieties of vegetables, pasture-raised Black Angus cattle, Tamworth and Berkshire pigs, an apple orchard, honeybees, and shiitake mushrooms. He invites the public to his seasonal farm stand, gives farm tours, and, in the fall, makes cider on his restored 1895 cider press. In 2009 Art added a windmill and solar-powered well. "The bees get more attention than anything," Art said. "People just love them."

640 Mariposa Road, Stanley (Gaston County), 704-263-2635, www.appleorchardfarmnc.com. Farm stand open June to September. Tours year-round by appointment.

With a Glean in Their Eyes

On larger farms, a fair amount of food is left in the fields after the professional pickers come through. Enter the volunteer gleaners with the Society of St. Andrew (www.endhunger .org), a Virginia-based nonprofit hunger relief ministry whose volunteers remove the leftovers from fields and packing houses.

The North Carolina branch, based in Durham and open since 1992, organizes some 11,000 volunteers yearly and gleans a whopping 18.6 million servings of fruits and vegetables, distributed to more than 2,600 charitable agencies statewide. And they can always use more.

Lewis Farm

You'll find Lewis Farm improbably tucked away in a suburban neighborhood that once was all farmland. On the November afternoon we visited, a group from Gastonia was having a birthday party here for the third year. The dozen children were all smiles as they scrambled over hay bales, checked out the stables, visited the chickens, and took turns horseback riding. Leading the horses around the ring were owners Robbie and Cathy Lewis, who share their forty-acre working family farm with the public. Make sure to ask Robbie to show you his collection of horse buggies. He made them all, by hand.

330 Lewis Road, Gastonia (Gaston County), 704-842-1208, www.lewisfarm.org. Occasional public events. Visits by appointment.

Lineberger's Maple Springs Farm

Harold and Patsy Lineberger moved to his family's farm thirty minutes west of Charlotte in 1984, but they'd already been selling U-pick strawberries in the area, making them one of the state's longest-running U-pick berry growers. Now the couple oversees two farms in two counties near Charlotte, the fifty-five-acre Lineberger's Maple Springs Farm in Dallas, which is the base for most activities, and the thirty-acre Berry Hill Farm in Iron Station, where they grow berries and fruit and have another U-pick strawberry patch. In the fall, Patsy, a retired teacher, leads hayride tours and is always slipping snippets of agricultural education into the children's activities.

906 Dallas-Stanley Highway, Dallas (Gaston County), 704-922-8688, maplespringsfarm.home.mindspring.com/. Open May to October, except August.

Stowe Dairy Farms

Tim and Gwen Stowe are on a mission to preserve not only the Stowe family's third-generation farm west of Charlotte but also its heritage. The couple, who moved into the attractive 1916 farmhouse atop a hill in 1988, are both educators. "Tim really has a passion for educating people about the early 1900s farming," Gwen said. To that end, he has restored a horse-drawn wagon, a seed separator, a hit-and-miss engine, and, his pride and joy, a Delco light plant, used in the early 1900s for generating electricity. On their eighty-acre portion of the larger Stowe Dairy Farms, the couple grows and sells produce and meat from their grass-fed black Angus cows. In the fall they turn their sorghum crop into syrup using an 1896 mill and outdoor furnace. They also raise and sell natural popcorn, and in the winter they run a choose-and-cut Christmas tree business, with cedar, cypress, and pine trees. Though the farm is steeped in tradition, it had a thoroughly modern moment in 2003 as the setting for a print advertisement unveiling the second-generation Prius.

169 Stowe Dairy Road, Gastonia (Gaston County), 704-228-9826, www.stowedairyfarms.com. Produce stand open in summer, "Sorghum Saturdays" in October, tree sales late November to December. Tours by appointment.

Carrigan Farms

Like many farms that have turned to agritourism, Carrigan Farms has U-pick strawberries in the spring, apple and pumpkin picking in the fall, and hayrides. During October, it opens "Forbidden Farms," a haunted trail through the woods "for those who dare the scare." Sometimes it sells produce, too. But the thirty-acre farm a half-hour north of Charlotte has something quite unique — its own beach. To reach this tropical paradise, open only by appointment, you drive past cornfields, an apple orchard, and more fields. After walking down a little path, you enter another world, gazing down at blue water and up at the surrounding granite cliffs of a quarry. To one side is a sandy beach, stairs down into the swimming hole, and a sitting area. A snack bar and changing area are nearby, making this a perfect if improbable spot for parties, weddings, and children's gatherings. Next time we're nearby, we're packing our swimsuits.

1261 Oak Ridge Farm Highway, Mooresville (Iredell County), 704-664-1450, www.carriganfarms.com. Produce sales May to October.
Quarry open March to November.

Mills Garden Herb Farm

"I started out doing crafty things with herbs, then got into culinary herbs, then I realized, I can make my own medicine," said Madge Eggena of Mills Garden Herb Farm. After moving from Charleston onto a ten-acre farm in Statesville in 1998, she finally had the opportunity to plant the culinary and medicinal herb garden she'd dreamed of. Madge organically farms several plots of herbs, which she sells at four farmers' markets, and also created a woodland garden filled with medicinal plants. In 2005, Madge and her sister, herbalist Jane Abe of Asheville, teamed up to offer cooking and medicinal classes at Mills Garden, which also include garden tours and herb gathering.

732 Mills Garden Road, Statesville (Iredell County), 704-873-3361, www.millsgardenherbfarm.com. Sales, tours, and classes by appointment.

Grateful Growers Farm

Considering everything they raise and grow, you'd think Grateful Growers Farm would cover tens of acres. Instead, it's only ten. "We try to grow food every place we can," said Natalie Veres, who with her partner, Cassie Parsons, has turned this farm just west of Lake Norman into a sustainable showcase of heritage livestock. The breeds they help preserve include their specialty, Tamworth hogs, along with Delaware chickens and Blue Swedish and Khaki Campbell ducks. You'll find their meat on the menus of most Charlotte-area restaurants that serve local meat and on sale at several stores and farmers' markets. In 2009 the pair started an all-local breakfast and lunch truck service called Harvest Moon Grille, which visits various locations around Charlotte. The farmers occasionally host on-farm dining events and have done a laudable job of fostering community among farmers, chefs, and customers.

3006 Mack Ballard Road, Denver (Lincoln County), 828-234-5182, www.ggfarm.com. Sales and tours by appointment.

Cassie Parsons (left) and Natalie Veres prepare for deliveries of their heritage-breed pork from Grateful Growers Farm in Lincoln County. Photo by Diane Daniel.

Historic Latta Plantation

For a look at early-nineteenth-century farm life, visit Historic Latta Plantation, a former cotton plantation turned into a well-maintained living-history farm on the grounds of the Latta Plantation Nature Preserve. The restored two-story plantation farmhouse, furnished with Federal period pieces, was home to merchant James Latta. Also on the grounds are a restored barn, log cabin, and smoke house. Many programs here teach children about 1800s farm life, but of course they'll be most drawn to the menagerie, which includes sheep, a donkey, mule, horse, cow, and hog. Tours of the home are guided, tours of the grounds are self-guided, and special events draw visitors throughout the year.

5225 Sample Road, Huntersville (Mecklenburg County), 704-875-2312, www.lattaplantation.org. Open daily.

Rural Hill

One of Mecklenburg County's best-kept secrets is Rural Hill, a gorgeous 265-acre working farm and home of the nonprofit Center of Scottish Heritage. Situated thirty minutes northwest of Charlotte, the former late-1700s plantation was the homestead of Major John and Violet Davidson. Visitors can follow a printed walking tour to see original and recreated buildings and enjoy miles of walking trails around the estate. A full-time farm manager oversees the small herd of Highland cattle (sold to breeders), hay production, and gardens. In September and October, a pumpkin patch, hayrides, and a six-acre corn maze attract families. November features a weekend of sheepdog trials. And, in its role as Scottish ambassador, the center in April sponsors the popular Loch Norman Highland Games, where kilted clans dance and bagpipes fill the air.

4431 Neck Road, Huntersville (Mecklenburg County), 704-875-3113, www.ruralhill.net. Public events throughout the year. Visits and tours by appointment.

Fisher Farms

When Dane Fisher met his wife-to-be, Maria, he asked her, "Do you like gardening?" "Yeah, gardening's great," she answered, not expecting he was talking about eight acres' worth. Though Maria doesn't have a farming background like Dane does, she has a master's degree in plant nutrition. He has a PhD in plant breeding. At their Fisher Farms, northeast of Charlotte, on land that has been in the Fisher family since 1933, the couple cultivates a variety of naturally grown vegetables. Their specialty is tomatoes, mostly heirloom varieties that they grow from seed, which they sell at farmers' markets. "We save seeds and come up with our own varieties," said Maria, who is the taste tester. Farm tours here come as an invitation to get your hands dirty. A little work will get you free produce and a look at how farming is really done.

7725 Stokes Ferry Road, Salisbury (Rowan County), 704-239-1719, www.fisherfarms1933.com. Sales and tours by appointment.

Laughing Owl Farm

Laughing Owl is possibly the Charlotte area's best-known farm thanks to farmer Dean Mullis's column in the *Charlotte Observer*. Called "Life on the Farm," it's taken from longer posts on his humorous blog. Dean and his wife, Jenifer, also are well known on the farmers' market circuit and have been active in farm education through their popular CSA and their membership in Slow Food Charlotte. On seven acres surrounding their house, the family grows a variety of produce, with potatoes and garlic being specialties, said Dean, who started the farm in 1989. They also sell lots of eggs, broilers from spring to fall, and turkeys for Thanksgiving. Part of Dean's reason for writing about farming is concern for its future. "We'd love to teach more people to do this," he said.

28016 Ryan Road, Richfield (Stanly County), 704-463-1555, www.laughingowlfarm.com. Sales and tours by appointment.

Dean and Jenifer Mullis take a break from the heat during a long day of work at Laughing Owl Farm in Stanly County. Photo by Diane Daniel.

Aw Shucks!

When Bonnie Griffin taught elementary school, she realized most children had no experience with farming. Aw Shucks!, her impressive seasonal agritourism enterprise, grew from her desire to "get kids out of the classroom and bring them into the farm." Since 2004 Bonnie has run a five-acre corn maze and a pumpkin patch and continues to add attractions on her twenty-nine acres, which include a haunted trail, a fishing pond, farm animals, toy duck races, and a turkey shoot. She loves "old-timey things" and has gathered artifacts over the years, from old farming equipment to an antique pickup truck. Her most attention-getting acquisition, in 2009, is what she has been told is the oldest remaining wooden train car on the East Coast, dating back to the 1800s. The farm's focal point is a very cute "General Store," where Bonnie sells old-fashioned candy, gifts, and goofy novelty toys for kids and kids at heart.

3718 Plyler Mill Road, Monroe (Union County), 704-709-7000,
www.awshucksfarms.com. Open September to November or by appointment.

Poplar Ridge Farm

For Poplar Ridge Farm owners Marianne Battistone and Philip Norwood, running a farm is an extension of their commitment to health and organic food. Both are athletes, and Marianne writes about health for national magazines and works in injury prevention part-time in New York City. They bought their beautifully maintained eighty-acre farm in 1994 when Philip started working in Charlotte. Farm managers oversee the day-to-day operations. While most of the bounty from their five acres of certified organic produce and flowers goes to their CSA subscribers, the surplus is sold to the public from the farm one afternoon a week. Several times a year, Poplar Ridge hosts chef's dinners, cooking classes, and "Farm Days," featuring a market and tours.

1619 Waxhaw–Indian Trail Road South, Waxhaw (Union County), 704-843-5744, www.poplarridgefarmnc.com. Open to public one day a week and during events.

FARM STANDS AND U-PICKS

Barbee Farms

After spending summers during high school selling surplus vegetables from his family's Barbee Farms, Brent Barbee joined the business full-time in 2008. Now his job has grown from selling from a small table at one market to managing produce sales at no fewer than five farmers' markets a week. Barbee Farms, owned by Brent's parents, Tommy and Anna, also operates a seasonal stand at the farm and sells produce to restaurants and grocers. "I'm the first generation of the past three to work at the farm full time," said Brent, who manages seventy acres of produce, fruit trees, and rotating crops. The farm scored a victory in 2009 when the State Department of Transportation altered roadway plans that would have severely disrupted the farmland and likely destroyed the family's long-standing home.

1000 Shelton Road, Concord (Cabarrus County), 980-521-1395, www.barbeefarms.net. Farm stand open April to October.

Yesterways Farm

When we arrived at Jane Biggers's farm, the longtime farmer was briefing a woman and her young daughter about how she does business at Yesterways Farm, northeast of Charlotte. "We offer some of the best food that can be produced, and we expect cooperation from our customers. That saying that the customer is always right doesn't apply here." We had to chuckle as the customer drove off, either thrilled or terrified, we couldn't tell. "I needed to tell her like it is," said Jane, a former trucker. "Our eggs are some of the best you'll ever have. The beef, pork and chicken are all sold from the fridges in our little shack. We use the honor system. If she didn't leave money, that's between her and the Lord." Jane, who runs the farm with her daughter, also grinds her own organic yellow cornmeal on an eight-inch gristmill and makes her own soap, both of which she sells at her small farmstand.

15401 Short Cut Road, Gold Hill (Cabarrus County), 704-279-5859. Open daily.

Knob Creek Farms

Jeff Crotts tends to the seventy-five acres of apple trees his father first planted south of Hickory in 1948. From there, Knob Creek Farms kept growing, adding thirty acres of peaches, thirty acres of blackberries, and eight acres of strawberries. All except the peaches have a U-pick component. In 1999 Crotts opened an airy gift shop and a homemade ice cream counter, called Knob Creek Farms and Creamery. Already picked fruit is sold here, as are dried apple snacks and other edible and decorative gifts.

6471 Fallston Road, Lawndale (Cleveland County),
704-538-1405. Open April to Christmas.

Patterson Farm

With a farm market, 350 acres of tomatoes, and twenty-five acres of strawberries, Patterson Farm has a thriving U-pick and retail operation. Over the years, many family members connected to this third-generation business thirty miles north of Charlotte have returned from other careers to keep the farm going. Agritourism has become a big part of its future. The farm hosts special events year-round and leads tours for schools, community groups, and individuals. Fall is the busiest time on the farm, when hayrides take kids through Patterson's forty acres of pumpkins. For Christmas, the market fills up with tens of thousands of poinsettias.

10390 Caldwell Road, Mount Ulla (Rowan County), 704-797-0013,
www.pattersonfarminc.com. Open spring to Christmas.

A customer sorts through green beans at the Charlotte Regional Farmers Market (Mecklenburg County). Photo by Diane Daniel.

FARMERS' MARKETS

Charlotte Regional Farmers Market

Charlotte Regional Farmers Market, open since 1984, has a strong and organized group of local growers, many of whom farm organically or naturally. Clustered together in Building B on Saturdays, they identify themselves with signs supplied by Slow Food Charlotte reading "Local Farmer, Local Food," distinguishing their stands, they hope, from nonfarmers selling produce from other farms near and far. (The market itself doesn't make that distinction.) The Charlotte market has expanded several times, most recently adding a craft building in 2005. Also on the grounds are two retail produce buildings, one wholesale, and a shed for greenery sales.

1801 Yorkmont Road, Charlotte (Mecklenburg County), 704-357-1269, www.agr.state.nc.us/markets/. Open daily.

The Davidson Farmer's Market

The Davidson Farmer's Market, in this small college town north of Charlotte, started in 2008 with twelve vendors, a number that had almost doubled within a month. By the next year, it was holding steady with about forty vendors, including farmers, food artisans, and craftspeople. The market was a grassroots effort partly inspired by the success of the long-running Matthews market to the south. Both have much in common: they feature only local farmers and products, are located in charming historic downtown areas, and harvest strong community support.

216 South Main Street, Davidson (Mecklenburg County), www.davidsonfarmersmarket.org. Held Saturday mornings May to October.

Matthews Community Farmers' Market

We were sorry to have just missed the action when we arrived at the Matthews Community Farmers' Market on Tomato Tasting Day, but we heard it was yummy. That event came on the heels of Canning and Preserving Day and Peach Crepe Fundraiser Day. The Matthews Community Farmers' Market in downtown Matthews, just south of Charlotte, has the enviable combination of a small-town feel with uptown events, including the popular "Ask a Chef" weekly series. Operating since 1991, the market is billed as the largest growers' market in the Charlotte area, as all vendors grow their own produce within fifty miles of Matthews. In the sweet little Farmers' Market Community House, a century-old building that was once part of a nearby cotton gin operation, shoppers can buy cold drinks and T-shirts and chat with Pauline Wood, market manager since 1995. Next on the lineup the week after we were there: Veggie Art Day. We must time our visits better.

105 North Trade Street, Matthews (Mecklenburg County), 704-821-6430, www.matthewsfarmersmarket.com. Held Saturday mornings April to November, otherwise on alternating Saturdays.

CHOOSE-AND-CUT CHRISTMAS TREES

Santa's Forest

John Herter's family has been farming in Lincolnton, north of Charlotte, for more than a century, but they started their Santa's Forest choose-and-cut tree business much later, selling their first Leyland cypress and Carolina sapphires to holiday shoppers in 2004. The farm already grew the trees for its Japanese-maple nursery business. Santa's Forest is run by John and his mother, Gale, but the whole family pitches in at Christmastime. Being farmers, not only do they offer hayrides and have animals on hand, they've also added such kid-friendly attractions as goat-milking and sheep-herding demonstrations.

4071 Herter Road, Lincolnton (Catawba County), 828-428-3774, www.japanesemaple.net/santaforest.htm. Open daily late November to December.

Helms Christmas Tree Farm

Henry Helms started Helms Christmas Tree Farm, south of Hickory, in 1976. Over the years he has planted eight acres of red cedar, white pine, Leyland cypress, Carolina sapphire, and several varieties of spruce. Unlike most choose-and-cut farms, Helms will sell any tree "bagged and balled," meaning it can be replanted. Helms also carries a small number of precut Fraser firs from the mountains. "Ten to one, my customers prefer Leyland cypress over Fraser," he said. His wife, Trudy, has a following for her homemade wreaths. Customers are taken into the forty acres of fields on a small covered wagon, a little different from the conventional hayride, Helms notes. The food sold here also is out of the ordinary—ham biscuits and barbeque, along with burgers and dogs. In July and August, customers can visit the Helms's acre of U-pick blueberries. If you can't make it then, Trudy sells her blueberry jam at the Christmas shop.

6345 Christmas Tree Lane, Vale (Lincoln County), 704-276-1835, www.helmschristmastreefarm.com. Open late November to December.

VINEYARDS AND WINERIES

WoodMill Winery

Shortly after Larry Cagle opened WoodMill Winery in 2006, he said, "I had brides lining up to get married here." And that was even before he added a large pavilion extending off the deck of the tasting room. Indeed, the rural setting amid rolling hills and rows of vines is inviting and unique to this area south of Hickory. Larry, a former research engineer in the nuclear industry, said he was compelled to open a winery as a way to entice his father, who had heart problems, to drink more wine. Dad was interested only in the sweet stuff, and so Larry bucked the regional trend and

Their Fifteen Minutes of Fame

North Carolina farms and dwellings have been in the spotlight in films, commercials, and radio shows. Here are some of the vehicles for their starring roles.

Blood Done Sign My Name: Two Cleveland County farms play parts in the 2009 film version of the book written by Tim Tyson, in which a black man is allegedly murdered by a local white businessman in Oxford, North Carolina. The cattle farm Twin Chimneys and Spake Strawberry Farm, both near Shelby, had roles in the film, while uptown Shelby filled in as downtown Oxford.

"Can't Tell Me Nothing": The 2007 music video to hip-hop artist Kanye West's single of the same name features comedian and Wilkesboro native Zach Galifianakis and folk singer-song-writer Will Oldham. The unscripted low-budget clip was shot at Zach's family farm in Wilkesboro and features the stars lip-synching to the song while acting crazy with farm equipment.

"Five Farms: Stories from American Farm Families": The multimedia project was coproduced by the Center for Documentary Studies at Duke University and aired on public radio in 2009. Farmers in the spotlight included Eddie Wise, a fourth-generation minority farmer at Wise Family Farm in Whitakers, north of Rocky Mount. (www.fivefarms.org)

Moving Midway: This documentary, released in 2007, was written and directed by Raleigh native Godfrey Cheshire, a Manhattan film-critic-turned-director. It follows the past and future of "Midway Plantation," his ancestral house the family relocated in Knightdale. The film addresses issues of slavery and race, as well as encroaching development. (www.moving midway.com)

Prius advertisement: The year was 2003 and the project was top secret. Commercial photographers set up shop for a week at Stowe Dairy Farms in Gastonia to photograph the second-generation Toyota Prius for a print ad that ran later that year in several national magazines, showing the farm's rolling pastures in the background.

Secret Life of Bees: Though set in South Carolina, the 2008 Hollywood film starring Dakota Fanning, Jennifer Hudson, and Queen Latifah was shot in North Carolina, mostly in Robeson and Pender counties. The pink farmhouse where many scenes took place is in Watha (500 Watha Road), while the peach orchard shots were filmed at Herring Farms outside of Lumberton, home to the roadside stand Geraldine's Peaches and Produce. Jacksonville-area berry farmer Julian Wooten coached the actors on beekeeping.

planted muscadine grapes instead of European varieties. He did later bow to the dry-red-wine crowd by adding a cabernet franc. On his fifty-two acres, he grows ten varieties of grapes. WoodMill also makes fruit wine with blueberries and blackberries, which Larry gets from local farms. As for the name WoodMill, Larry had planned to open a wood shop on one level, but the winery proved so popular that he didn't have the room, or the time.

1350 Woodmill Winery Lane, Vale (Lincoln County), 704-736-7733, www.woodmillwinery.com.

Dennis Vineyards Winery

Situated on fourteen acres along a quiet road outside of Albemarle, Dennis Vineyards Winery feels both sophisticated and laid back. This family operation started with muscadine wines made by Pritchard Dennis, following in the steps of his father. In 1997 Pritchard and his son, Sandon, started the winery, which has ten acres in cultivation. A few years later, they opened a gift shop and tasting room, and Sandon's wife, Amy, joined the business full time. Dennis Vineyards has been especially successful with its semidry muscadines, and also is known for its wide varieties of fruit wines. In 2006 the winery added a large events facility across the street, which has become wedding central. Called A Place in the Vineyard, its rather standard exterior belies the striking interior, featuring vaulted ceilings and wooden floors. Many couples marry in the gazebo down the hill, amid rows of lush vines, providing a sweet beginning.

24043 Endy Road, Albemarle (Stanly County), 800-230-1743, 704-982-6090, www.dennisvineyards.com.

Uwharrie Vineyards

From its well-manicured lawn bordered by crape myrtles to the jazz fusion piped into the tasting room and patio, Uwharrie Vineyards, outside of Albemarle, has a more uptown than down-home feel. Forty-five acres of grapes are planted on Uwharrie's eighty-four-acre property, with more in the works. While the setting was once rural, it's now more suburban. Winemakers Chad Andrews and owner David Braswell make both vinifera and muscadine wines. The winery and large gift shop opened in 2005, and later a 4,000-square-foot banquet hall was added for weddings and other events. Along with the outside seating and a few picnic tables, customers are welcome to spread a blanket on the lawn.

28030 Austin Road, Albemarle (Stanly County), 704-982-9463, www.uwharrievineyards.com.

STORES

The Bradford Store

Kim Bradford greets half the customers at The Bradford Store by name and everyone else with a big "Hey!" She and her husband, Grier, have restored what was in the early to mid-1900s the community's gathering place. Their store focuses on local products and produce, including many items from the couple's own gardens, planted nearly up to the front steps. The Bradfords restored the building, which had belonged to Grier's great-grandfather and had been closed since 1958, reopening it in 2005. Soon, residents from Lake Norman and North Charlotte were making it a regular stop. "We wanted it to have a similar feel from when it was a general store," said Kim. Along with their own produce, the Griers carry milk from Homeland Creamery in Julian, Goat Lady Dairy cheese from Climax, apples from North Carolina orchards, and more. Separate blacksmith and organic garden supply shops share the twenty-three acres, making this an even more interesting place to stop and shop, or just say hey.

15915 Davidson-Concord Road, Huntersville (Mecklenburg County), 704-439-4303, www.bradfordstore.com.

Renfrow Hardware and General Merchandise

If your shopping list goes something like this: collard seeds, two live turkeys, and a pressure cooker, you need to make only one stop — Renfrow Hardware and General Merchandise in downtown Matthews. Farmers have been coming to this sprawling old-fashioned store since 1897, said David Blackley, who bought Renfrow Hardware in 1984 after having gone there almost daily as a kid. In the spring the store carries live poultry, including chickens, turkeys, and guinea hens. Organics supplies have been available here for decades and are selling more than ever, as is canning and pickling equipment. "We've been just overwhelmed by the response from the food people the last few years," said David. His flora sales are totally food focused. "We are real big in vegetables, herbs, berries, and perennials that are edible, and we've always sold bulk seeds," David said. "If it grows and you can eat it, we try to sell it."

188 North Trade Street, Matthews (Mecklenburg County), 704-847-4089.

DINING

Artisan

One of the three culinary education programs in the area has a student-run restaurant, the white-tablecloth Artisan, which opened in 2004 in the International Culinary School, part of the Art Institute of Charlotte. The menu changes with the growing seasons, as students rely on local farms for much of Artisan's produce. Chicken, pork, and seafood also are bought from small North Carolina farmers and fishermen, giving the students a firsthand look at a farm-to-table operation. Because the popular forty-six-seat restaurant has very limited hours and serves only lunch, reservations are strongly recommended. In lieu of tips, diners pay a 15-percent fee that goes toward the culinary school's scholarship fund.

2110 Water Ridge Parkway, Charlotte (Mecklenburg County), 704-357-5900, www.artinstitutes.edu/charlotte. $-$$

A Fresh Approach to Food Distribution

Chefs have a few ways of working with farms to get produce and protein on the menu week after week. Sometimes they deal directly with farmers, even to the point of coordinating plantings or animal purchases in advance to coincide with seasonal menus. It's also not unusual for small farmers to deliver goods to restaurants or for chefs to shop at the farmers' market, but that takes more time and money. A fairly new option is a local-farm-products distributor.

The state's leading small-farm produce distributor is Pittsboro-based Eastern Carolina Organics (ECO; www.easterncarolinaorganics.com), which works only with organic farms. It started in 2004 as a project of the Carolina Farm Stewardship Association. Now about twenty farmers share ownership. ECO markets and delivers certified organic produce from more than forty growers to an ever growing number of customers, including restaurants, corporate cafeterias, grocery stores, and distributors.

New River Organic Growers (www.newriverorganicgrowers.org) is a cooperative of small organic and sustainable farmers in northern mountain counties. It delivers members' produce, eggs, and meat to restaurants in the region, including ones in the towns of Boone, Blowing Rock, and West Jefferson.

The Charlotte food-service market is the first beneficiary of Farmers Fresh Market (www.farmersfreshmarket.org), an Internet-based produce market that links Charlotte-area chefs with small farmers and ranchers in Rutherford County to the west. It also handles delivery. The service, started in 2007 by Foothills Connect Business and Technology Center, a nonprofit rural economic development program, hopes to grow into other regions. Farmers Fresh also runs a seasonal farmers' market on Saturdays in Forest City.

Out west, Mountain Food Products, located at the Western North Carolina Farmers Market in Asheville, has offered chefs and stores one-stop shopping by buying and distributing local products since 1984. In 2010, the state got its first pasture-based meat distribution service when the Center for Environmental Farming Systems launched Farmhand Foods.

Barrington's Restaurant

Since Bruce Moffett opened the forty-five-seat Barrington's Restaurant in 2000, named after his hometown in Rhode Island, the sophisticated spot has stayed among the best of Charlotte's restaurants. The menu changes seasonally and features local and organic produce and meats year-round, including pork from Grateful Growers and organic produce from New Town Farms. It seemed that every farmer we spoke with had done business with Barrington's, and Bruce often appears at farm-to-fork and farmers' market events. The restaurant describes its simple and inventive food as "upscale American with a homey twist."

7822 Fairview Road, Charlotte (Mecklenburg County), 704-364-5755, www.barringtonsrestaurant.com. $$–$$$

Flatiron Kitchen and Tap House

A collective sigh of relief was heard when chef Tim Groody announced his new venture in 2010. When Groody, a veteran and leader of Charlotte's local-food scene, left a longtime chef's post in 2009, local diners found themselves without one of their favorite cooks. The following summer, he and two partners opened Flatiron Kitchen and Tap House in nearby Davidson. "While we serve wine, there's more emphasis on beer, with simple, fresh, playful bar food. It's nice that we have a great farmer's market right across the street," he said of the popular Davidson Farmer's Market. "It's the perfect spot for us."

215 South Main Street, Davidson (Mecklenburg County), 704-237-3359. www.flatirononmain.com. $$

Global

Chef Bernard Brunet of Global takes diners around the world with his French-inspired dishes, many of them starting out in North Carolina soil. "Eating a locally grown vegetable compared to something you get in a grocery store is like driving a Mercedes after a Ford," said the French native. Bernard opened the sixty-seat restaurant in 2006, in the Ballantyne neighborhood of Charlotte. When we spoke with him late fall, he was lamenting the end of the heirloom tomato season. "We make heirloom risotto, gazpacho with heirlooms, heirloom salad." Bernard sources almost all his produce and mushrooms locally, and meat based on availability, including chicken and pork. "I'm really trying to promote local food, with a French flair." Mais oui.

3520 Toringdon Way, Charlotte (Mecklenburg County), 704-248-0866, www.global-restaurant.com. $$–$$$

Rooster's Wood-Fired Kitchen

Six years after opening the upscale Noble's Restaurant in Charlotte, Jim Noble in 2006 debuted a more casual, affordable eatery that immediately became one of the hottest spots in town. The menu at Rooster's Wood-Fired Kitchen is completely à la carte, with offerings including local cheeses, cured meats, and of course grilled chicken. But what really excites Jim, who works with a long list of farmers, is the ability to offer so many vegetable sides, especially during harvest seasons. "In the summer, we may have four different bean dishes alone. Beans and peas are one of my new passions. I sit down with farmers and try to get them to try different kinds. I just love all the varieties and local names for them."

6601 Morrison Boulevard, Charlotte (Mecklenburg County),
704-366-8688, www.roosterskitchen.com. $$

The Inn at New Town Farms

Well-known long-time organic farmers Melinda and Sammy Koenigsberg decided in 2009 to open part of their farm, south of Charlotte, to diners and guests, calling it The Inn at New Town Farms. "We want to encourage the connection between farm and food," Melinda said. While the lodging part was still being worked out when we visited, dining plans were going full steam ahead. First, the couple transformed the lovely 5,000-square-foot home that had been occupied by Sammy's late father into a rural retreat. It sits on New Town Farms' twenty-nine acres, though trees hide most of the farm activities. Visiting chefs cook up masterpieces for private parties, and the Koenigsbergs host farm dinners that are open to the public.

4512 New Town Road, Waxhaw (Union County), 704-843-5182,
704-534-5582. Prices vary with function.

RECIPES

Asian Lettuce-Leaf Wraps with Spicy Dipping Sauce

Mary Roberts at Windcrest Farm recommended these healthy wraps in one of her informative farm newsletters (sign up at www.windcrestorganics.com). The organic farm in Union County sells plants and produce at Charlotte-area farmers' markets.

SERVES 4 AS AN APPETIZER

1	pound ground pork
1	carrot, finely chopped, about ½ cup
¼	pound shiitake mushrooms, chopped
2	tablespoons ginger, finely chopped
1	jalapeño pepper, finely chopped
1	clove garlic, finely chopped
1	tablespoon Asian fish sauce
1	tablespoon Chinese cooking wine (Shaoxing) or dry sherry
	Kosher salt
1	tablespoon peanut or vegetable oil
1	tablespoon fresh cilantro, chopped
20	bib or romaine lettuce leaves (if using romaine lettuce remove the thick part of the rib)
1	bulb kohlrabi, julienned
1	bunch scallions, thinly sliced

In a large bowl, combine the pork, carrot, shiitake mushrooms, ginger, jalapeño, garlic, fish sauce, and cooking wine. Season with salt.

Heat the oil in a large skillet or wok over high heat. Add the pork mixture and stir-fry, breaking up the pork, until it is cooked through and beginning to brown, about 6 to 8 minutes. Remove from heat and stir in the cilantro.

Serve the pork next to the stacked lettuce leaves on a large platter. To eat, spoon the pork into the lettuce leaves, top with some kohlrabi, scallions, and sauce (recipe below), and roll up.

2   tablespoons sugar

2   tablespoons water

1   tablespoon soy sauce

1   tablespoon rice wine vinegar

1   tablespoon freshly squeezed lemon juice

¼   teaspoon sesame oil

1–2   teaspoons ground chili paste

In a small bowl combine the sugar and water and stir to dissolve. Add the soy sauce, vinegar, lemon juice, sesame oil, and chili paste. Stir to combine.

Chipotle-Roasted Pork with Homemade Tortillas

Grateful Growers Farm in Lincoln County is well known in the Charlotte area for its Tamworth hogs, a heritage breed. Grateful Growers co-owner Cassie Parsons, a trained chef, puts the spotlight on the pork with this marinated masterpiece.

SERVES 4 TO 6

1   tablespoon brown sugar

1   teaspoon paprika

1   teaspoon dry mustard

2   teaspoons toasted cumin seed, ground

3   tablespoons olive oil

4   pounds bone-in pork shoulder

    Kosher salt and freshly ground black pepper

    Juice and zest of 2 fresh large lemons, about ½ cup

    Juice and zest of 2 fresh oranges, about ½ cup

3   canned chipotle peppers in adobo, chopped,

      plus 2 tablespoons of the sauce

2   cloves garlic minced, about 2 teaspoons

In a small bowl mix together the brown sugar, paprika, dry mustard, cumin seed, and olive oil. Rub the mixture all over the meat, sprinkle with salt and pepper, and set in a large bowl or pan.

Meanwhile, in a large bowl add the lemon juice and zest, orange juice and zest, chipotle peppers and sauce, and garlic. Mix well. Pour the citrus/pepper sauce over the roast, covering it well. Cover the meat with plastic wrap and refrigerate for at least 2 hours, preferably overnight. Turn the roast halfway though the marinating.

Preheat the oven to 250 degrees.

Bring the pork to room temperature. Place the pork and marinade in a roasting pan, cover with foil, and cook for about 4 to 5 hours or until the meat is so tender it will fall off the bone. Remove from the oven and let rest for 15 minutes before pulling apart. Pour the pan juices into a heat-proof bowl. Skim off the fat and pour into a small saucepan, place over medium heat, and reduce by ⅓. Add the sauce to the pork and serve in homemade tortillas (recipe below) with roasted poblanos and home-made salsa.

MAKES 16 TORTILLAS

HOMEMADE TORTILLAS

3   cups unbleached flour
2   teaspoons baking powder
1   teaspoon salt
6   tablespoons vegetable shortening or lard
    About 1 cup warm water

Mix the dry ingredients in a bowl and add the vegetable shortening or lard. Cut the shortening into the dry ingredients with a pastry blender, two knives, or your fingers until it resembles coarse meal. Add the warm water, a little at a time, until the dough is soft but not sticky. Knead the dough with your hands for a few minutes until soft and pliable.

Break off small (1- to 2-inch diameter) balls of dough and let them rest for about 10 minutes (longer is okay). Preheat a cast-iron skillet over medium heat. On a lightly floured surface, use a rolling pin to roll each ball into a thin 6-inch round. Cook the tortilla in the hot skillet until golden-brown speckles appear on the dough. Flip to the other side and finish cooking. Wrap the tortilla in a clean kitchen towel to keep warm. Repeat with the remaining tortillas.

Chocolate Shortcake with Strawberries, Cream, and Grand Marnier

Bruce Moffett's Barrington's Restaurant in Charlotte, featuring elegant farm-fresh dishes, consistently appears atop best-of lists. In this modern twist on old-fashioned strawberry shortcake, Bruce takes in-season strawberries and elevates them to a royal status befitting the Queen City.

SERVES 8

2	cups all-purpose flour
2	tablespoons cocoa powder
¼	cup plus 1 tablespoon sugar
1	tablespoon plus ½ teaspoon baking powder
6	tablespoons cold unsalted butter, cut into small pieces
¾	cup heavy cream
2	hard-boiled egg yolks, mashed
2	tablespoons unsalted butter, melted
3	cups ripe strawberries, hulled and quartered
2	teaspoons granulated sugar
2	tablespoons Grand Marnier
1	cup mascarpone
2	tablespoons powdered sugar
	Zest of 1 orange
	Juice of half an orange
1	cup whipped cream (about ½ cup whipping cream), whipped to soft peaks

Preheat oven to 350 degrees.

Lightly butter a baking sheet. Sift the flour, cocoa, ¼ cup of the sugar, and the baking powder into a medium bowl. Stir to combine. Add the cold butter and, using a pastry blender or your fingertips, cut in the butter until it resembles coarse meal. Add the cream and egg yolks and stir with a fork until the dough is just moistened and holds together. If necessary, add more cream a tablespoon at a time.

Turn the dough out onto a lightly floured surface and knead until it forms a smooth dough, about 3 to 4 times. Do not overwork. Pat or roll the dough ¾-inch thick. Using a floured 3-inch cookie cutter, cut out 4 rounds of dough. Gather up the dough scraps, reroll, and cut out the remaining 4 rounds. Transfer to the prepared baking sheet.

Brush the biscuits with the melted butter and sprinkle with the remaining 1 tablespoon of sugar. Bake on the middle rack of the oven for 12 to 15 minutes until the biscuits are firm to the touch. Transfer the biscuits to a rack and let cool.

Meanwhile, in a medium bowl, combine the strawberries, sugar, and Grand Marnier and allow to sit at room temperature for about an hour or longer if needed. In a separate bowl stir together the mascarpone, powdered sugar, orange zest, and juice. Gently fold in the whipped cream. Strain the juice from the berries and reserve. Mix the berries with the mascarpone.

Just before serving, cut the shortbreads in half horizontally. Place the bottom half on the plate. Spoon on the strawberry mascarpone mixture and top with the other shortbread half. Drizzle the reserved liquid around the dessert and serve.

Rockingham

Surry
Elkin

Stokes
• King

Reidsville
•

• East Bend
Yadkin —
Yadkinville
•

Winston-Salem
•

Guilford

Forsyth
Davie —
Mocksville
•

High Point
•

Greensboro
•

Davidson

Asheboro
•

Randolph

Triad

Beyond Greensboro, Winston-Salem, and High Point, the nine counties of the Triad are largely rural, from the foothills of the Brushy Mountains to the rolling plains to the south. The Triad is known for its large textile, tobacco, and furniture corporations, including R. J. Reynolds Tobacco Company, based in Winston-Salem. The decline of tobacco spurred a new industry here—viniculture and enology, or grape growing and wine making. In the early 1970s, the late Jack Kroustalis of Westbend Vineyards in Forsyth County planted European grapes against the advice of everyone. By 2010 the state is expected to have more than 100 wineries, and even more vineyards, with the majority in the Triad. They draw visitors from around the state and beyond, who come to sit, sip, and even stay a few nights.

FARMS

Denton FarmPark

A fund-raiser for his local rescue squad in 1970 started Brown Loflin on a journey that is still going strong. Over the years at what is now known as Denton FarmPark, in Denton, southwest of Asheboro, the Loflin family has collected fifteen restored buildings from the nineteenth and twentieth centuries. Structures on the 140 acres include a general store, grist mill, log cabin, barn, church, smokehouse, blacksmith shop, and plantation home. They run five events a year, drawings tens of thousands of visitors. The most popular is the Southeast Old Threshers' Reunion, one of the largest antique farm-machine shows in the Southeast.

1072 Cranford Road, Denton (Davidson County), 336-859-2755, www.threshers.com, www.countrychristmastrain.com.

Ijames Heritage Farm

Ijames Heritage Farm has been operating since 1792, but owners Todd and Reba McInnis didn't open it up to visitors until 2008. They give tours of the farm, which Reba said covers "thousands of acres" and also rent it for parties and other events. While the McInnises both work "off farm," they grow some crops and keep laying hens. The land, thirty minutes southwest of Winston-Salem and only three miles off Interstate 40, is situated along a quiet country road in a pastoral setting. Farm tours include a look at goats, donkeys, chickens, and demonstration tobacco and cotton patches.

367 Sheffield Road, Mocksville (Davie County), 336-492-7529, 336-407-2900. Tours by appointment.

The Edible Schoolyard at Greensboro Children's Museum

The Greensboro Children's Museum in 2010 proudly became the first museum in the country to build an Edible Schoolyard. The much heralded educational program was initiated in 1995 at a school in Berkeley, California, by Alice Waters, local-food guru, chef, author, and owner of Chez Panisse restaurant. Her aim is to integrate gardening and cooking into schools' academic curricula in the hope of transforming the way children eat. The Schoolyard in Greensboro, which the public can tour, is on a half-acre of the museum's outdoor play area and includes organic vegetables, herbs, fruits, flowers, trees, and shrubs with winding natural walkways and seating areas. It also holds a teaching kitchen, classrooms, a chicken coop, arbors, a green house, and composting and recycling stations.

220 North Church Street, Greensboro (Guilford County), 336-574-2898, www.gcmuseum.com.

Steeple Hill Farm

While Steeple Hill Farm has been known for its horses for decades, in 2008 owner Renee Weidel started to turn her picturesque 140-acre spread into an October farm adventure for youngsters. The farm also is available for private rentals and overnight stays. "Everything is geared toward children twelve and under. We don't do anything remotely scary," said Renee, who has owned the farm, thirty minutes northwest of Greensboro, since 1983. On hand for the fun are miniature horses, boar goats (which Renee breeds), ducks, chickens, pigs, and a flock of sheep another farmer raises on her pasture. "The thing so staggering to me is the reaction of the children," said Renee, who recounted their squeals of delight, especially when watching pigs eat the food the kids toss them through a feeding chute. Other thrills include an activity-filled hayride, a hay maze, and a pumpkin patch.

7000 Belford Road, Summerfield (Guilford County), 336-643-4090, www.steeplehillfarm.com. Fall events from late September to November. Lodging $$

Caraway Alpacas

Pondering ways to save and use farmland that had been in the family for nine generations, Bobby and Anne Poole and their daughter and son-in-law, Teresa and Mike Johnson, first considered aquaculture. "While flipping through one of the farm magazines we had for research, I saw alpacas, and that was that," Teresa said. In 1997 the family opened one of the state's first alpaca farms, Caraway Alpacas. Today they take care of about thirty alpacas, whose pasture covers fifteen of the family's eighty-five rolling acres outside of Asheboro. "I love their calmness and tranquility. You kind of get addicted to them," said Teresa. Her parents, using the animals' fiber, weave and crochet beautiful shawls, scarves, and other items, which are for sale in their alpaca-themed shop. In case you're wondering, her apricot-colored alpacas are actually white. "That color is from rolling around in the red clay," Teresa said.

1079 Jarvis Miller Road, Asheboro (Randolph County), 336-629-6767, www.carawayalpaca.com. Sales and tours by appointment.

Goat Lady Dairy

Don't spend too much time around Steve Tate at Goat Lady Dairy unless you want to raise dairy goats, make cheese, grow produce, tend to turkeys and chickens, and host gourmet dinners. The public face of this sixty-acre family enterprise is so passionate about sustainable farming you'll want to practice whatever he's preaching. It's no surprise that Tate is an ordained minister who used to work as a counselor. "We are the new American farmer: food with a face," he told a crowd at one of the dairy's popular dinners and tour. "There really are easier ways to make a living, but not as fun." Steve and his wife, Lee, moved here from Minnesota in 1995 to join his sister, Ginnie, the "goat lady," in starting the dairy. When it opened in 1996, Goat Lady was one of the state's first farmstead cheese makers, and it set the bar high. Their cheese is now sold at several farmers' markets and

The Top of Their Class

North Carolina has had some long-standing agricultural education programs, with a handful more starting in the past decade. Colleges and universities also are offering more food studies courses and programs.

The state land-grant colleges, North Carolina State University and North Carolina A & T State University, have their roots in conventional agricultural training and are known for their undergraduate and graduate degree programs, from business management to soil science.

The private Warren Wilson College, which opened in 1894 as the Asheville Farm School for youths and became a four-year college in 1967, offers a four-year degree in environmental studies that focuses on sustainable agriculture.

Since 1991, Appalachian State University (ASU) has offered a bachelor's degree in sustainable development, with a focus on agriculture. Later ASU added viticulture courses, and in 2008 it started the Appalachian Center for Mountain Winegrowing, an outreach program to develop the vineyard and winery industry in the mountains. The university also plans to offer an undergraduate degree in viticulture.

Surry Community College in Dobson started a viticulture program in 2000 that has helped launch many Yadkin Valley wineries. Students work at a five-acre vineyard and Surry Cellars, their bonded winery on the college campus. In 2010 the school opened the Shelton-Badgett North Carolina Center for Viticulture and Enology, a $5 million complex that includes a teaching lab, winery, classrooms, and special events hall.

In 2001 Central Carolina Community College in Pittsboro became the first community college to offer an associates degree in sustainable agriculture. The program now draws students from other states.

In 2009, after running a horticulture program for decades, Western Piedmont Community College in Morganton added a sustainable agriculture program.

For early starters, the Arthur Morgan School near Burnsville is a sustainable farm-focused boarding school for seventh through ninth graders.

grocers. Several thousand fans attend their two yearly Open Farm Days, held in conjunction with neighboring Rising Meadow Farm.

3515 Jess Hackett Road, Climax (Randolph County), 336-824-2163,
www.goatladydairy.com. Two open-farm days.
Otherwise, sales, dinners, and tours by appointment.

Rising Meadow Farm

Looking to get out of their careers in education but still do something educational, hobby farmers Ann and Ron Fay, along with their daughter, Beth, and son-in-law, Winfield Henry, decided to buy a farm. They moved to Rising Meadow Farm, southeast of Greensboro, in 2003. After major restoration work on the late-1800s farmhouse, the Fays first ran a bed and breakfast, then turned to farming full time. (Beth and Winfield live in a separate home on the 128 acres.) The couple tends to about 100 sheep for wool and meat, and keeps a few other animals, including guard llamas. Rising Meadow holds three public events yearly. They join their neighbor Goat Lady Dairy for a spring and fall Open Farm Days, while Shearing Day is held in February, featuring a professional sheep shearer. "I sell half the fleeces on that one day," Ann said. "It's pretty amazing."

3750 Williams Dairy Road, Liberty (Randolph County), 336-622-1795,
www.risingmeadow.com. Two open-farm days.
Otherwise sales and tours by appointment.

Century Farm Orchards

Century Farm Orchards is as picturesque a place as you'll see. Rows of apple trees flank a white farmhouse surrounded by rolling meadows, all situated along a quiet country lane. David Vernon moved to his family's 250-acre farm in 1999 and started a small apple nursery featuring old-timey southern apples from the 1600s to the early 1900s. He also has some trees from the family farm's beginnings, in 1872. David turned to famed Chatham County apple preservationist and historian Lee Calhoun for advice on grafting and growing. When Lee decided to sell his nursery, David, a chemistry teacher at West Alamance High School, stepped in to fill the gap. David has grafted about 500 varieties of apple trees, as well as ten heritage pear varieties, and he planted a peach orchard. Apple varieties include Roxbury russet, the oldest American apple, originating in Massachusetts in the 1600s, and yellow June, in existence since 1845. Of yellow June, he writes in his catalog, "as a child I loved this apple because it was the first to ripen at my grandparents' farm." Thanks to him, both the farm and the apples live on.

1614 Rice Road, Reidsville (Rockingham County), 336-349-5709,
www.centuryfarmorchards.com. Sales by mail order and appointment.
Tours by appointment. Open Saturdays in November only.

Cornerstone Garlic Farm

While less than an acre of garlic may not sound like much, Natalie and Steve Foster harvest upward of 40,000 heads of garlic a year on family land that was once a tobacco farm. What started in 1996 as a way for Natalie, a former teacher, to have a summer income became her full-time job five years later. From its first crop of elephant garlic, which is actually a leek, Cornerstone has expanded to ten varieties. The Fosters, regulars at the Greensboro Farmers' Curb Market, also have added shallots, shiitake mushrooms, vegetables, and blackberries to their organically grown offerings. Natalie makes and sells garlic braids and a line of garlic and herb seasonings and dip mixes. Her latest offering is garlic-herb dog biscuits. Better to have garlic breath than Fido breath.

1249 Tate Road, Reidsville (Rockingham County), 336-349-5106.
Sales and tours by appointment.

Hills and Hollows Farm and Museum

Guerrant and Janet Tredway started out buying a few llamas for their own enjoyment in 2005, "and it's just sort of grown from there," said Guerrant, who goes by "G. A." In 2009 the couple started to open their 100-acre farm, west of Eden and near the Virginia border, to visitors. Not only can people touch and feed llamas and see how their fiber is processed, they're invited to a fascinating show. G. A. will put a llama through an obstacle course, an activity for which some of their fifteen or so animals have won medals. G. A. and Janet are avid collectors of country and farm antiques, including household items, toys, and tools, which they present in two buildings, one loosely set up as a general store. G. A. also restored a sharecropper's cabin. "We want people to see how things used to be," said G. A., who grew up on the family land, used first for a dairy and then for a tobacco farm. "If anybody wants to donate anything, just let us know."

5007 Price Road, Stoneville (Rockingham County), 336-573-2993,
www.hillsandhollowsfarm.com. Tours by appointment.

His llamas come running when G. A. Tredway is ready with their feed at Hills and Hollows Farm and Museum in Rockingham County. Photo by Diane Daniel.

Pennwood Farm

On the early-spring afternoon we visited Pennwood Farm, a trio of three-day-old goats following farmer Anna Micciulla bounded over to greet us. "I'm bottle feeding them, so they follow me everywhere," she said. Anna's Plan B for the farm turned out to be an A-plus idea. In 2005, she, with husband, Bob, and daughter, Xiomara, in tow, fled corporate life in New England and returned to the Piedmont, where she'd grown up in a dairy-farming family. Her original idea was to make farmstead goat cheese on their eight-acre spread, but that proved too pricey. The problem was she'd already purchased a herd of Nubian goats. So instead, Anna created Pennwood Puritanicals, which makes goat-milk soap and other products. In the lineup are bar soap, liquid soap, lotion and cream, and herbal body care products, including lip balm and herbal bath teas, which she sells at the Piedmont Triad Farmers Market. Along with the goats, the family tends to chickens, sheep, turkeys, and ducks.

138 Carroll Farm Road, Reidsville (Rockingham County), 336-280-3529, www.pennwoodpuritanicals.com. Sales and tours by appointment.

Horne Creek Living Historical Farm

Horne Creek Living Historical Farm, south of Pilot Mountain State Park, offers an outstanding look at the state's rural and agricultural heritage from about 1900 to 1910. Included on the 104 acres of this state historic site, once the home of two family farms, are a farmhouse, well house, fruit house, smokehouse, tobacco barn, and corncrib. All are within a quarter-acre of the new visitor center, completed in 2009. The walking trail to the buildings passes the family cemetery, gardens with heirloom vegetables, and pastures containing a small number of heritage livestock. In 1997 the Southern Heritage Apple Orchard was established with a gift of 400 varieties of heritage apples grafted and grown by Lee Calhoun, an apple preservationist from Chatham County. Special events are held here throughout the year, with the most popular being the fall "Cornshucking Frolic," which draws up to 10,000 frolickers.

308 Horne Creek Farm Road, Pinnacle (Stokes County), 336-325-2298, www.nchistoricsites.org/horne/horne.htm.

Keep Your Fork Farm

From their modest brick ranch home in King, thirty minutes north of Winston-Salem, Jane Morgan Smith and her husband, Rick, are raising what they hope will be a fortune-yielding supply of truffles. The underground mushrooms, sought by epicures, are worth about $800 a pound. But the finicky fungus, which grows on roots of trees, is not a sure thing. In a decade, the Smiths have harvested ten pounds — more than most anyone else in the state. They have 725 inoculated filbert and oak trees planted on under two acres. Their pet dogs have been trained to sniff for the underground gold during harvest season, in the winter. Though Jane has been promoting North Carolina truffles for years, it took a visit from domestic diva Martha Stewart to get people's attention. "Most people who come on a farm tour are interested in growing truffles themselves, but some want to come just because Martha Stewart was here."

1194 Marshall Smith Road, King (Stokes County), 336-631-8080, www.keepyourforkfarm.com. Tours by appointment.

FARM STANDS AND U-PICKS

SandyCreek Farm

Lexington residents Brenda and John Garner didn't set out to be farmers. It started with the two of them helping Brenda's mother with the large garden her father had planted on a fifteen-acre parcel of family land outside of town. Eventually they built a home there, and in 2002 SandyCreek Farm was born. Now they tend to the garden and the muscadine vines that Brenda's father had planted, as well as heirloom pear and fig trees donated by John's parents. In 2006 the couple started growing organic shiitake mushrooms, and they now have about 700 inoculated logs, which visitors may view. In 2009 the couple opened a cheery farm store and gift shop on the mostly wooded lot, where they sell Brenda's homemade jams and jellies, dried shiitakes, shiitake dip mix, produce in season, and some local crafts. "I had to do something with all the grapes," said Brenda of her jam making.

3160 South Highway 150, Lexington (Davidson County), 336-250-6702,
www.sandycreekfarm150.com.

Clod-Buster Farms

One thing Zane Sells knew when he opened his own farm in 1997 after years of working with his grandfather was that it wouldn't bear his name. "It would sound like a real estate office," he said. He came up with Clod-Buster Farms, named for the clumps of dirt that farmers are always having to bust up. The former tobacco farmer diversified in 2004, starting with four acres of U-pick strawberries, then adding produce and a five-acre corn maze.

5500 Leonard Farm Road, Kernersville (Forsyth County), 336-409-0796,
www.clodbusterfarms.com. Stand open April to November.
Maze open September to November.

Early Farms

In 2002 fifth-generation Early Farms transitioned from tobacco to produce, which is sold out of their tiny farm stand between Greensboro and Burlington. A variety of produce is grown, along with two acres of U-pick strawberries. The family also runs an eight-acre corn maze and other fall attractions on family land a mile from the farm. Called J Razz and Tazz

Of Pests and Pesticides

There is considerable debate in this country and elsewhere about pesticide use. You can assume that most North Carolina farms, unless they are organic, biodynamic, or market themselves as sustainable or pesticide- and chemical-free, use some amount of pesticides.

Typically, the more important the appearance of the commodity and the shorter its growing season (such as fruits and vegetables), the more often it is sprayed with pesticides.

The use of some pesticides requires training and protective equipment. Farmers who use these "restricted-use pesticides" have to be certified by the state to ensure they are trained to follow proper procedures for safely applying pesticides and harvesting treated crops.

Most states, including North Carolina, do not require mandatory reporting on pesticide use, though farms are required to keep records the state can check if performing a selected pesticide inspection. That falls under the Structural Pest Control and Pesticide Division in the Department of Agriculture. Also, surveys to determine pesticide use in some commodities are conducted by the department's Agricultural Statistics Office. These results are posted online.

Toxic Free NC (www.toxicfreenc.org) believes that state and national pesticide programs are not thorough or well-funded and that toxic pesticides are overused in North Carolina and elsewhere, harming food, water, and people, particularly children and farmworkers. The Raleigh-based nonprofit group advocates for safer alternatives.

Farm's Corn Maze, the nicely decorated area contains several attractions, including a hayride, a haunted hayride, and a play area for young children.

6031 Bethel Church Road, Gibsonville (Guilford County), 336-697-2473.
Open April to September. Corn maze at 466 Peeden Drive, Gibsonville.
Open September to October.

Ingram's Strawberry Farm

In 1979 Richard and Kathryn Ingram planted six acres of strawberries on their forty rolling acres of land to help save the farm that has been in their family since the 1800s. Now son Dean and daughter-in-law, Rhonda, run Ingram's Strawberry Farm, along with their three daughters. One of them, Casie, took over the kitchen operation at age sixteen, selling jams and her own baked goods from their farm stand. The berry fields are spread around the farm, which is divided by a pond with a pier and a seating area. In 2009 the Ingrams started to bring in goats and calves for viewing. "Now the kids are entertained while their moms pick berries," Rhonda said.

6121 Riverside Drive, High Point (Guilford County), 336-431-2369,
www.ingramfarm.com. Open April to July.

Rudd Farm

When Kenneth and Joan Rudd opened an acre of U-pick strawberries in Greensboro in 2001, "we were scared to death that no one would come," Joan said. Now, at nine acres of berries and counting, they're wondering how many more acres will be enough to satisfy the ever growing crowds. Kenneth and Joan's son, Matt, represents the fourth generation of Rudd Farm, which stopped growing tobacco in 2004. Two years later, Joan retired from her textile-industry job and shepherded the opening of a produce stand. From there, the Rudds sell a wide variety of vegetables grown on some thirty acres, as well as greenhouse tomatoes in the spring. As for the strawberries, which are usually ready mid-April, try to come during the week, Joan recommended. "Weekends are just crazy."

4021 Hicone Road, Greensboro (Guilford County), 336-621-1264,
www.ruddfarm.com. Open April to October.

Millstone Creek Orchards

Byrd and Diane Isom bought eighty-five acres between Asheboro and Siler City in 2002 and planted a few apple trees. "Then it just kind of kept growing," said their daughter, Beverly Mooney, who comes up from Florida several times a year to help out at her parents' Millstone Creek Orchards. The Isoms now have fifteen acres in fruit, including ten varieties of apples, blueberries, blackberries, grapes, and peaches, all U-pick and they-pick. Fall visitors also can pull pumpkins out of the one-acre patch. From the kitchen adjoining the orchard's small gift shop and snack bar, Diane churns out apple butter, cider, jams, and baked goods. Out back, picnic tables and tire swings in a shady spot invite customers to linger. Events, which all cater to families with young children, include a fall festival, a Halloween hayride, and group tours. "This really started because my father needed something to do when he retired," said Beverly, who plans to eventually help run the farm. "How did he know it would be something I'd be so happy doing, too?"

506 Parks Crossroads Church Road, Ramseur (Randolph County), 336-824-5263, www.millstonecreekorchards.com. Open June to November.

Haight Orchards

Eileen and Ed Haight cashed out of their family orchard in pricey Westchester County, New York, blaming high taxes and a push toward development, and bought an abandoned tobacco farm outside of Reidsville in 1987. When they married in 1961, Eileen was a city girl. "I knew nothing about farming," she said. That changed fast, and she joined Ed, a sixth-generation fruit grower, in the orchard business. Over the years they've added peach trees, and they now have four acres of peaches, five of apples (with ten varieties), and about 100 nectarine trees. The beautifully maintained orchard, situated on rolling hills, is a lovely place to spend an afternoon, and warmer than New York.

2229 Pannel Road, Reidsville (Rockingham County), 336-427-6933, www.haightorchards.com. Open July to September.

Tuttle's Berry and Vegetable Farm

For generations, the Tuttle family has farmed a few hundred acres in the Shiloh community southwest of Eden. Patriarch D. L. Tuttle, who died in 2002, was a prime mover in getting the North Carolina General Assembly to fund a state-run farmers' market in the Greensboro region. The family continues to farm, and it sells its produce, strawberries, blueberries, and peaches at the state market and at its own attractive roadside stand. In 2009 the Tuttles opened a six-acre corn maze near the farm stand. "It's been a lot of work," said Helen Tuttle, widow of D.L., "but we're excited about it. The kids get to see animals, take a hayride through the woods, and go down to the spring."

2701 Highway 135, Stoneville (Rockingham County), 336-627-5666, www.tuttlefarms.com. Farm stand open April to November, Monday through Saturday. Corn maze and hayrides late September though October.

FARMERS' MARKETS

Greensboro Farmers' Curb Market
All is quiet in downtown Greensboro on a summer Saturday morning until you get near the Greensboro Farmers' Curb Market. By 8:00 A.M., the place is hopping. Dating to 1874, the Greensboro market is one of the state's oldest and most popular. Farmers come from North Carolina and bordering states. For better and worse, more than 100 vendors and their customers are packed into an old National Guard Armory building, with some overflow outside. While the bustle adds to the festive mood, shopping can be a full-body-contact experience, but we suppose that's part of the fun.

501 Yanceyville Street, Greensboro (Guilford County), 336-574-3547. Held Saturday mornings year-round and Wednesday mornings May to December.

Piedmont Triad Farmers Market
Many farmers rank Piedmont Triad Farmers Market as tops among the five state-run farmers' markets for local produce. Opened in 1995, the market includes two buildings with about seventy vendors selling plants and produce. Saturday is the best day to catch the smaller, sustainable farmers, most of whom sell in Building E. Unlike at some state markets, farmers here must grow and sell their own products, and market managers do check up on them, as the farmers report proudly.

2914 Sandy Ridge Road, Colfax (Guilford County), 336-605-9157, www.agr.state.nc.us. Open daily.

Rockingham County Farmers' Market
Business has been booming at the Rockingham County Farmers' Market since it opened in 2004 in the former log-built stables of the Chinqua Penn estate. The building, once owned and used by the Penn family in the early

to mid-1900s, was beautifully restored. Vendors from a nine-county area sell produce and meat from their farms at the market, as well as berries, eggs, wines, herbs, and cheese. A craft gallery is housed in the former tack room. Brenda Sutton, also known as the Produce Lady, is one of the market organizers and a vendor as well. Brenda is also the county extension director and has made a series of videos on market food preparation. As she said, "No farmer wants to return home from the market with bushels of unsold eggplant after customers pass it by because they have no clue how to prepare it."

1944 Wentworth Street, Reidsville (Rockingham County), 336-342-8230, www.co.rockingham.nc.us/farmark.htm. Held Saturday mornings and Wednesday afternoons May to October.

Winston-Salem Downtown Farmer's Market and Krankies Local Market

Downtown Winston-Salem is home to two markets, a small one that started in 2009 and focuses on sustainably or organically grown offerings, and a more traditional market, open since 1994. The longer-running Winston-Salem Downtown Farmer's Market meets under a pavilion tucked inside a block of downtown. About twenty-five vendors carry produce, flowers, baked goods, plants, and crafts. Krankies Local Market is held inside Krankies Coffee, a downtown café and gathering spot housed in an old meat-packing warehouse. All the vendors at this rapidly growing market, organized by Krankies and the Triad Buying Co-op, grow or make their own products using sustainable or organic methods.

City market on Sixth Street between Cherry and Trade, Winston-Salem (Forsyth County), 336-354-1500, www.dwsp.org. Held Tuesday and Thursday mornings May to September. Krankies, 211 East Third Street, 336-722-3016, www.krankiescoffee.com. Held Tuesday mornings May to November.

VINEYARDS AND WINERIES

RayLen Vineyards

The long, winding drive through RayLen Vineyards to its signature tasting room evokes wineries in Napa Valley. If not for the hum of traffic outside, you'd be hard pressed to believe RayLen is just off Interstate 40, near Mocksville. The winery, opened by Joe and Joyce Neely in 2001 in a former

North Carolina is home to about 100 wineries and even more vineyards.
Photo by Selina Kok.

dairy, was an early arrival in the Yadkin Valley. It now grows a dozen varieties of grapes on 40 of its 115 acres. The tasting room, filled with award-winning wines, is furnished with chairs made from wine barrels, which they also sell from their store. In 2009 it led the pack in another category by becoming the first winery to run partly on solar energy.

3577 Highway 158, Mocksville (Davie County), 336-998-3100, www.raylenvineyards.com.

Westbend Vineyards

In 1972, against the advice of many experts, Jack and Lillian Kroustalis planted European grapes instead of southern muscadine varieties on seventeen acres in Forsyth County. A decade later, the trailblazing West-bend Vineyards had doubled the acreage (which now totals sixty) and sold grapes to several regional wineries. As Jack and Lillian moved into wine-making in the 1990s, winning several national competitions, Westbend quickly gained a reputation for quality. Jack's death in 2006 drew tributes nationwide from people who recognized his contribution to wine making in the South. Lillian and other family members have continued the operations. In 2008 Lillian renovated the 1850s farmhouse across the street from the winery. The home's iconic image has been used on several West-bend labels for decades.

5394 Williams Road, Lewisville (Forsyth County), 866-901-5032, 336-945-5032, www.westbendvineyards.com.

Carolina Heritage Vineyards

In 2009 Carolina Heritage Vineyards opened its doors as the first organic vineyard and winery in the state. Owners Clyde and Pat Colwell established the thirty-five-acre farm just east of Elkin after retiring from careers in education and technology, respectively. "When Clyde asked me, 'How about we start a vineyard and winery?,' I said, knowing how much vineyards are sprayed, 'If we go organic, I'm game,'" Pat said. All their grapes are either native muscadines or disease-resistant French hybrids. The couple pulls weeds by hand and also counts on their hungry chickens and guinea hens to help control the insect population. On many Saturday afternoons you'll find a bluegrass jam on the front porch of the tasting room, while Pat has the inside stocked with crafts from local artisans. "There's just so much talent around here that I want everyone to experience it."

170 Heritage Vines Way, Elkin (Surry County), 336-366-3301, www.carolinaheritagevineyards.com. Closed January.

Elkin Creek Vineyard

When Elkin native Mark Greene bought the 1896 Elkin Mill along Elkin Creek and surrounding land, it was to rescue the mill. But then he caught the Surry County wine bug and built a vineyard on his sixty acres. "We do everything by hand here," explained Mark, owner and winemaker at Elkin Creek Vineyard. He and family members not only cleared the land and planted the grapes, they built a windmill to pump water for irrigation and constructed a beautiful post-and-beam winery and restaurant, The Kitchen at Elkin Creek. Last we checked, the restaurant was open only for special events.

318 Elkin Creek Mill Road, Elkin (Surry County), 336-526-5119, www.elkincreekvineyard.com.

Grassy Creek Vineyard and Winery

Grassy Creek Vineyard and Winery, once a dairy barn and creamery known as Klondike Farms, was started in 2006 by two couples who bought 200 acres of land, planted vines, and opened a now popular winery. The wine is made in the former 10,000-square-foot dairy barn, and the tasting room, fashionably decorated in shabby-chic style, is housed in the old horse stable. While Grassy Creek produces several dry whites and reds, it's best known for its sweet wines packaged in 750-milliliter milk bottles, a nod to Klondike's once renowned chocolate milk. Another piece of the past remains in the Klondike Cabins on the grounds, which the winery rents out. They were once part of the rural retreat owned by John Hanes of Hanes Hosiery.

235 Chatham Cottage Circle, State Road (Surry County), 336-835-2458, www.grassycreekvineyard.com. Lodging $$

McRitchie Winery and Ciderworks

Sean McRitchie of McRitchie Winery and Ciderworks, arguably the state's most versatile winemaker, has helped start many Yadkin Valley wineries and is winemaker at a few of them. Sean worked in West Coast and Australian vineyards before landing in the South a decade ago to help launch Shelton Vineyards. His father, Robert, who put Oregon wineries on the map, later followed to run the viticulture program at Surry County Community College. Sean and his wife, Patricia, opened their own small, casual tasting room overlooking rolling hills in 2007. Along with several award-winning vintages, the winery produces a tangy hard cider from heritage Pink Lady apples, picked just up the road in the Brushy Mountains.

315 Thurmond P.O. Road, Thurmond (Surry County),
336-874-3003, www.mcritchiewine.com.

Round Peak Vineyards

Since buying Round Peak Vineyards in late 2008, Ken Gulaian and his wife, Kari Heerdt, have breathed new life into this winery, beautifully situated in the foothills of the Blue Ridge Mountains and ten miles west of Mount Airy. What hasn't changed at the winery, founded by two couples in 2005, is winemaker Sean McRitchie, one of the state's best regarded. Ken and Kari started a second label, Skull Camp, in 2009. They also added a host of mostly dog-friendly activities, including concerts, sunset tastings, and even a campout on the thirty-two-acre spread. They also opened the house next to the winery to overnight guests. Avid cyclists, Ken and Kari encourage fellow riders to park at the winery, go for a spin (they've mapped out routes), and of course return in time for happy hour.

765 Round Peak Church Road, Mount Airy (Surry County),
336-352-5595, www.roundpeak.com. Lodging $$

Shelton Vineyards

If you like your wineries more upscale than down-home, Shelton Vineyards, one of the state's best known and the inspiration to many new winemakers, is the perfect place. It is sophisticated and manicured, with a full-service restaurant (Harvest Grill), a state-of-the-art winery, a large gift shop and tasting room, and a spring-through-fall concert series under a band shell on the well-tended lawn. Since their early arrival on the viticulture scene in 2000, brothers, owners, and native sons Charlie and Ed Shelton have expanded from 60 to 200 acres of vineyards and sell more than a dozen European wine varietals. Shelton has won enough awards and fans to become known for producing some of the best wines in the state. Unofficially, locals know it's also one of the best places to take visiting relatives.

286 Cabernet Lane, Dobson (Surry County), 336-336-4724,
www.sheltonvineyards.com.

Stony Knoll Vineyards

We made the mistake of choosing a sweltering August day to bicycle to Stony Knoll Vineyards. Situated along a country road in rural Surry County, the scenery was plentiful but water-refill options were not. Finally we reached Stony Knoll's five acres of vineyards, on sloping land divided at the crest by scalloped lines, and up the hill we pedaled to the newly built tasting room. The moment we walked inside with bright-red faces, co-owner Van Coe rushed to get us cold water and implored us to sit a spell in the air-conditioned room. We did not refuse. In 2001, Van and his wife, Kathy, planted vineyards on part of their forty-eight-acre farm, which has been in her family since 1896. Van's brother-in-law, Lynn Crouse, is the winemaker and Van, a former banker, is the farmer, while Kathy and other family members work with customers. Since Stony Knoll opened in 2004, it also has become a popular wedding site. We can see why.

1143 Stony Knoll Road, Dobson (Surry County), 336-374-5752, www.stonyknollvineyards.com.

Raffaldini Vineyards

Instead of cultivating a down-home feel, like most wineries in the state do, Raffaldini Vineyards wants to transport its guests to Italy. Hence its tagline, "Chianti in the Carolinas." Raffaldini's founder, financier Jay Raffaldini, is a first-generation Italian American who, with other family members, started the vineyard in 2005. Although located just inside Wilkes County, the winemaker markets itself with neighboring vineyards in Yadkin County. The Raffaldinis planted French and Italian grape varieties, but the goal is to offer only Italian wines, such as sangiovese, barbera, pinot grigio, and of course chianti. In 2008 the winery's temporary tasting room was abandoned for a show-stopping villa on a beautifully landscaped hilltop. The unobstructed views of the Brushy and Blue Ridge mountains and over the vineyard's 100 acres are unparalleled, which is why you'll be hard-pressed to find an empty patio seat here on nice days.

450 Groce Road, Ronda (Wilkes County), 336-835-9463, www.raffaldini.com.

Divine Llama Vineyards

With dozens of wineries in the state, it doesn't hurt to offer something different. Perhaps the most unusual twist can be found at Divine Llama Vineyards in East Bend, which combines a llama farm with a winery. With only

From the patio at Raffaldini Vineyards (Wilkes County) in the Yadkin Valley, customers can enjoy sweeping views of the Brushy and Blue Ridge mountains. Photo by Selina Kok.

a parking lot between pasture and tasting room, visitors sipping their merlots can watch llamas cavort. "Ninety percent of people go to the pasture before coming to the tasting room," said co-owner Michael West. "The llamas are people magnets, for sure." West and his wife, Julia, co-own the farm (called Four Ladies and Me), while they and longtime friends Thomas and Julia Hughes own the winery. Together they bought seventy-seven acres in 2006, planted five acres of vinifera grapes, and opened the winery in 2009. The Wests and their three daughters have raised llamas since 2004, and they now keep a herd of about thirty-five. On most Saturdays and by appointment, the Wests will give winery customers a tour of the farm. During the grape harvest, the llamas are put to work, Michael said. "They wear packs with five-gallon buckets for the grapes, and we have fields full of volunteers who want to help them out."

4179 Divine Llama Lane, East Bend (Yadkin County), 336-699-2525, www.divinellamavineyards.com, www.fourladiesandme.com.

Flint Hill Vineyards

Alcohol sales are nothing new at Flint Hill Vineyards. Co-owner Tim Doub's grandfather, who was born in the home where the tasting room is located, used to run a distillery, selling jugs of whiskey and alcohol. Tim and his wife, Brenda, opened Flint Hill in 2005 after doing major renovations to the lovely 1870s yellow farmhouse Tim grew up in. The windmill in its logo is a nod to Brenda's Dutch heritage. Five acres of vinifera grapes surround the home, located on a lazy country road popular with bicyclists, who often congregate here. In 2007, the Doubs and chef Sean Wehr joined forces to open an outstanding and cozy fine-dining restaurant, called Century Kitchen.

2133 Flint Hill Road, East Bend (Yadkin County), 336-699-4455, www.flinthillvineyards.com. Dining $$

Hanover Park Vineyard

When Amy and Michael Helton created their winery in 1996, inspired by a trip to France, they started the wave of wineries in Yadkin County. Not only were the artists and former educators facing the uphill task of learning to grow grapes and make wine, they also tackled the renovation of the long vacant 1897 farmhouse on their twenty-three acres. The beautifully redone home is now the perfect stage for tasting the Heltons' dry reds and whites. For a touch of homespun southern France, try their *vin d'orange*, a rosé blended with oranges, lemons, vanilla beans, sugar, and brandy.

1927 Courtney-Huntsville Road, Yadkinville (Yadkin County), 336-463-2875, www.hanoverparkwines.com.

Laurel Gray Vineyards

From the farm pond to the porch glider outside the tasting room inside a renovated milking parlor, Laurel Gray Vineyards exudes a cheery, country feel that's punctuated by the red-and-white color scheme and beds of roses. On eighty-four acres that has been in Benny Myers' family for more than 200 years, he and his wife, Kim, planted their first vinifera vines in 2001. The operation has grown to include ten acres of vineyards and a full-scale winery, where Laurel Gray also crushes grapes for other winemakers. Make sure to check out Kim's artwork, which decorates not only the walls of the tasting room and gift shop but also Laurel Gray's labels.

5726 Old Highway 21 West, Hamptonville (Yadkin County), 336-468-9463, www.laurelgray.com.

RagApple Lassie Vineyards

You're likelier than not to find owners Lenna and Frank Hobson holding court at RagApple Lassie Vineyards, which was one of the first to open

Cindy Shore prepares produce to deliver to her CSA members and her Sanders Ridge Winery and Restaurant in Yadkin County. Photo by Diane Daniel.

in the Yadkin Valley, in 2002. RagApple, northwest of Winston-Salem, is named after the beloved champion calf Frank showed as a child. Frank is the quintessential farmer (RagApple is the state's only winery started by career farmers), while Lenna is the magic marketer. She devises all sorts of community-building events, including concerts in their natural amphitheater, to attract crowds. She's also a strong supporter of many local charities, which is why every event here includes charitable dona-tions. Lenna describes the winery's fetching logo as depicting "a Carolina moon turned right for good luck, Pilot Mountain, and a cow with her legs crossed at the ankle, like a proper southern lady."

3724 RagApple Lassie Lane, Boonville (Yadkin County),
866-724-2775, www.ragapplelassie.com.

Sanders Ridge Winery and Restaurant

Sanders Ridge Winery and Restaurant in Yadkin Valley, open since 2009, is a welcome addition to the state's short list of wineries with great res-taurants. The 350-acre century farm is at the center of operations. From there, fifth-generation farmer Neil Shore grows fifteen acres of vinifera and muscadine grapes and three acres of certified organic vegetables, with more acreage planned. Cindy Conti was brought in as farm manager, and the business partnership turned into a romantic one. Now married, Neil and Cindy Shore manage the farm, a CSA, vineyards, the winery, and a forty-seat restaurant. The drinking and eating takes place in an eye-catching post-and-beam structure built from lumber and stone harvested off Sanders Ridge Farm. The wooded setting, with decks outside overlooking a small pond and a stone fireplace inside, makes it perfect for all seasons. Starr Johnson, well known in the area for her regional cooking using fresh ingredients, runs the kitchen. "Whatever is in season from the farm we use in the restaurant," she said.

3200 Round Hill Road, Boonville (Yadkin County),
336-677-1700, www.sandersridge.com. $$

STORES

Homeland Creamery

Brothers Chris and David Bowman, representing the sixth generation of their family at Bowman Dairy in Julian, added the creamery in 2002. Since then, Homeland Creamery has been a popular spot to go for fresh-scooped, fresh-made ice cream and milkshakes, which can be enjoyed at picnic tables beside the shop. Ice cream is only part of the store's farm-fresh offerings. You can also pick up hormone-free white milk, chocolate milk, cream, hamburger, and, from other local farms, sausage and farm eggs. Homeland's products are sold in many stores as well.

6506 Bowman Dairy Road, Julian (Guilford County), 336-685-6455,
www.homelandcreamery.com. Tours by appointment.

DINING

Meridian Restaurant

Meridian Restaurant co-owner Trevor Dye holds a cooking demonstration from spring to fall at Krankies Local Market, teaching the public what he does regularly as a chef, using local sources to create a new menu daily. The upscale restaurant in downtown Winston-Salem opened in 2007 with a Mediterranean bent but has started to focus on seafood, Trevor said. "We use a vendor who sells seafood exclusive to North Carolina fishermen," he said.

411 South Marshall Street, Winston-Salem (Forsyth County),
336-722-8889, www.meridianws.com. $$–$$$

Noble's Grille

Noble's Grille has been a fine-dining favorite in Winston-Salem since it opened in 1992. Owner Jim Noble, who also operates restaurants in Charlotte, including Noble's Restaurant, opened his first dining spot in High Point in 1984, which he has since sold. "Back then you couldn't find local stuff anywhere, and you couldn't even buy fresh herbs, so I grew my own." Now Jim works with a variety of local farmers to supply his produce and meat.

380 Knollwood Street, Winston-Salem (Forsyth County),
336-777-8477, www.noblesgrille.com. $$–$$$

Bistro Sofia

From spring through early fall, produce gathering is simple for Beth Kizhnerman, owner and chef at Bistro Sofia. She just goes outside the kitchen door of her 120-seat restaurant and plucks items from the large urban garden, including lettuce, eggplant, squash, beets, and tomatoes. Just don't ask her how to grow it, Beth said, because she's not the gardener. Since opening in 1999, the elegant Bistro Sofia has remained one of Greensboro's top restaurants, much of this success due to Beth's attention to freshness. She buys local eggs, some meat, and, when her garden isn't producing, she shops at the farmers' market. In warm weather, diners flock to the patio enclosed by high red brick walls, across from Guilford College. Customers also may wander through the extensive vegetable and herb garden to see what might be on their dinner plates.

616 Dolley Madison Road, Greensboro (Guilford County),
336-855-1313, www.bistrosofia.com. $$–$$$

Lucky 32 Southern Kitchen

Greensboro diners have been lucky for a long time. A few years after the first Lucky 32 Southern Kitchen opened in 1989, then chef Bart Ortiz and crew were out looking for fresh produce. "We had what we called our 'field truck' that we'd do pick-ups in," recalled Bart, whose efforts helped spur the creation of Eastern Carolina Organics produce distributors. More than

Foodies Unite

In the 1980s Italian writer Carlo Petrini railed against McDonald's arrival in Rome, as well as the growing industrialized food system, which he said was destroying food varieties and flavors. He started what became the Slow Food movement, a backlash against fast food. Today, the Slow Food organization is worldwide. (It, in turn, has been the victim of a backlash, with detractors claiming elitism.)

Slow Food U.S.A. (www.slowfoodusa.org) oversees the nonprofit group's activities in North America, carried out through about 170 local chapters, with several in North Carolina. The group's mission is to offer educational events and public outreach that promote taste education, advocate sustainability and biodiversity, and connect consumers and producers (farmers and food artisans).

While Slow Food's reach is global, the Southern Foodways Alliance (www.southernfoodways.com), formed in 1999, is concerned with food cultures of the changing American South, from roadside barbecue shacks to fine restaurants. The alliance, an institute of the Center for the Study of Southern Culture at the University of Mississippi in Oxford, stages conferences on food culture, produces documentary films and oral histories, and publishes food-related essay collections. Members include chefs, farmers, academics, writers, and food lovers.

two decades later, Lucky 32, which also has a Cary location, continues to buy locally. "It's a big part of the ethos here," said chef Jay Pierce. "We're really trying to support the local economy." Lucky 32 shares some farmers with Print Works Bistro and Green Valley Grill, two other, pricier Greensboro establishments also owned by Quaintance-Weaver Restaurants and Hotels.

1421 Westover Terrace, Greensboro (Guilford County), 336-370-0707, www.lucky32.com. $–$$

Sticks and Stones

We wish more pizzerias were like Greensboro's Sticks and Stones, which became an instant hit when owner Neil Reitzel opened it in 2008. The salads are locally sourced, and so are the stone-fired pizza toppings. The sausage and beef are from Moore Farm in Randolph County, the dairy products are from Homeland Creamery, the goat cheese is from Goat Lady Dairy, the charcuterie is made on the other side of town, and the veggie toppings and herbs are farm fresh. And then there's the dough, prepared with locally milled organic flour from Lindley Mills. The focus on local and sustainable carries through to the interior, artfully designed with recycled wood and metal, and the to-go boxes, made with recycled paper.

2200 Walker Avenue, Greensboro (Guilford County), 336-275-0220, www.sticksandstonesclayoven.com. $–$$

Sweet Basil's

Working for commercial catering companies, Sweet Basil's owner Renee Schroeder wasn't happy about the many artificial ingredients they used. "I don't eat that way myself," she said. Her idea was to start her own catering company, but as she and her husband looked for space, they came across a century-old Quaker farmhouse and decided to open a restaurant there. That was 2007, and while she initially had to seek out farmers, "now farmers come to me," she said. She and her chef purchase produce and meat from some twenty farms and serve only East Coast wild-caught fish. "What I most love about working with local farms, when you get a tomato from a farmer, you know that tomato was loved."

620 Dolley Madison Road, Greensboro (Guilford County), 336-632-3070, www.sweetbasilsrestaurant.com. $$–$$$

SPECIAL EVENTS AND ACTIVITIES

Dixie Classic Fair

The Dixie Classic Fair, which started in downtown Winston-Salem in 1882, is the second-largest agricultural fair in North Carolina, after the State Fair in Raleigh. Now run by the city of Winston-Salem, it draws an average of 325,000 visitors a year and is located on seventy-seven acres north of downtown. Agricultural competitions abound, involving everything from decorated apples and potatoes to various breeds of cattle, sheep, poultry,

Thinking Outside the Lunchbox

The local-food movement and the alarming rates of child-hood obesity and diabetes have spurred parents, teachers, health officials, and politicians to rethink the food served in school cafeterias and work to get farm-fresh food into school meals.

The North Carolina Farm to School program started in 1997 by serving apples grown in Henderson County. Several dozen school districts are now part of the program, which includes local purchasing, nutrition, education, and farm tours.

Separately, in the western counties, the Appalachian Sustainable Agriculture Project runs its own farm-to-school program. Called Growing Minds, it works in six counties with farmers, educators, and communities to serve local food in schools while expanding opportunities for farm field trips, experiential nutrition education, and school gardens.

swine, and wine. A favorite attraction is Yesterday Village, a collection of nineteen log structures that were built in the 1800s, including a one-room church, log homes, and a general store and post office.

421 West Twenty-Seventh Street, Winston-Salem (Forsyth County), 336-727-2236, www.dcfair.org. Held in October.

North Carolina Wine Festival

The state's most popular wine festival is the North Carolina Wine Festival, which debuted in 2001. Produced by WSJS radio and held in Clemmons, just west of Winston-Salem, it draws more than forty wineries and 25,000 wine appreciators, many of them setting up tents in Tanglewood Park to form their own little shaded communities. Musicians perform throughout the day and vendors hawk food, arts and crafts, hot sauce, and even fortunes. In the future, we predict even more people will enjoy the North Carolina Wine Festival.

4061 Clemmons Road, Clemmons (Forsyth County), 336-777-3900, www.ncwinefestival.com. Held in May.

North Carolina Wild Foods Weekend

There are farmers and then there are foragers. They're both taking food out of the ground, but the foragers are finding it, not planting it. Every spring since 1975, folks from several Eastern states have gathered at the Betsy-Jeff Penn 4-H Educational Center in Reidsville to go wild together at the North Carolina Wild Foods Weekend. They learn about edible wild plants and animals, and the climax is a wild feast prepared from foods gathered by the participants. "We have about 200 wild dishes at our feast, with nine groups who cook with an experienced wild-foods cook," said organizer Debbie Midkiff of Durham. "We usually have venison, bear, rattlesnake, fish, and quail. Salads might have pike, dandelion, daylilies, redbud, ramps, nuts, and berries." Even the drinks are sourced in the wild, such as sassafras, spicebush, and peppermint tea, perfect for washing down the rattlesnake casserole.

804 Cedar Lane, Reidsville (Rockingham County), 919-489-2221, www.wildfoodadventures.com. Held in April.

Yadkin Valley Wine Festival

One of the best things about the Yadkin Valley Wine Festival, which started in 2002, is its location at Elkin Municipal Park near historic downtown Elkin, a gem of a small town. Of late, the festival, sponsored by Yadkin Valley Chamber of Commerce, has been attended by more than 10,000 people sampling from some two dozen Yadkin Valley wineries. Music, food, and craft vendors are on hand as well, making this quite a lively day at the park.

399 Highway 268 West, Elkin (Surry County), 336-526-1111, www.yvwf.com. Held in May.

Yadkin Valley Wine Tours

If you're the type who would rather sit back and let someone else do the driving, Yadkin Valley Wine Tours is for you. The tours, which started in 2005, visit wineries in the Yadkin Valley, and perhaps beyond in the future. During stops at three to four wineries, your guide (likely owner John Byrd) will impart information on the history of North Carolina wineries, as well as how to see, smell, swish, sip, and, yes, spit wine (when necessary). We liked the balance of wineries the day we visited: a small and large one in rural settings, another in an urban location, and the camaraderie with our fellow wine tourists.

336-793-4488, www.yadkinwinetours.com.

RECIPES

Corn Skillet Medley

Brenda Sutton, a.k.a. the Produce Lady and the director of the Rockingham County Extension Service, has made a series of videos showing simple ways home cooks can use what's available from North Carolina farmers' markets. (View them at www.theproducelady.org.) This is one of our summertime favorites.

SERVES 6

2	tablespoons olive oil
1	pound zucchini, julienned
1	clove garlic, finely chopped
1	pint cherry tomatoes, halved
2	ears corn, kernels removed
	Kosher salt and freshly ground black pepper
2	tablespoons fresh basil, chopped, or other herb of your choice

Heat the olive oil in a large skillet over medium heat. Add the zucchini and cook for 3 minutes, until it begins to lose its crunch. Add the garlic and cook for 1 minute more. Add the tomatoes and corn and cook for 4 more minutes, until the tomatoes have softened slightly and the corn is cooked through.

Season with salt and pepper and sprinkle with fresh basil before serving.

Garlic and Basil Cream Sauce with Shrimp

Natalie Foster of Cornerstone Garlic Farm in Rockingham County created this recipe for a garlic festival cook-off. Not surprisingly, she placed as a finalist. The rich dish "had the crowd licking the sample plate," she reported. After making this dish, we believe it.

SERVES 4

12	ounces penne or other pasta
1	pound medium shrimp, peeled
10	cloves of garlic (8 minced and 2 thinly sliced; hardneck garlic recommended)
4	tablespoons butter
1	medium shallot, minced

1 cup heavy cream

10–13 basil leaves (10–12 minced, 1 whole)

1 ½ cups freshly grated Parmesan

Salt and freshly ground black pepper

Cook the pasta as directed on the box. Drain.

In a small deep skillet, sauté the shrimp and 2 cloves of minced garlic in 1 tablespoon of butter until the shrimp are just cooked. Take the shrimp out of the pan and add the rest of the butter.

Add the remaining minced garlic and the shallots and sauté for 3 minutes over low heat. Add the cream, basil, 1 cup of the cheese, 2 cloves of sliced garlic, and salt and pepper to taste. Simmer over low heat, partially covered, for 7 to 10 minutes until thickened.

Return the shrimp to the pan, mix with the sauce, and then pour over the pasta. Top the dish with the rest of the cheese and a basil leaf.

Super Simple Strawberry Cobbler

Teenager and third-generation farmer Casie Ingram has been making treats for the farm stand at Ingram's Strawberry Farm in High Point since she was eleven. This easy-to-make cobbler, which she sells by the slice or the dish, is one of her customers' favorites.

SERVES 4 TO 6

1 cup whole milk

1 cup self-rising flour

1 ¼ cup sugar

Pinch salt

4 cups strawberries, sliced

Preheat oven to 350 degrees.

In a medium bowl, combine the milk, flour, 1 cup of the sugar, and salt. Stir to combine. In another medium bowl combine the strawberries and remaining ¼ cup sugar and stir until the sugar dissolves.

Pour ½ of the milk/flour mixture into the bottom of a 5-cup baking dish. Add the fruit and top with the remaining milk/flour mixture. Place in the oven to bake for 50 to 60 minutes, until the top is golden brown.

Southern Collard Greens

"These are different; you should try them," Lucky 32 Southern Kitchen chef Jay Pierce advises the collard-averse about his flavorful treatment of this much-maligned green. He's right. You, too, will become a collard convert.

SERVES 4

1	pound collard greens
¼	pound pork fatback, rinsed well and cut into 1-inch cubes
1	yellow onion, sliced ¼-inch thick
2	medium carrots, sliced (about 1 ½ cup)
1	ham hock
5	cups chicken broth
¼	cup apple cider vinegar
2	tablespoons Worcestershire sauce
2	tablespoons soy sauce
1	teaspoon dried thyme
½	teaspoon freshly ground black pepper

Pick through the collards and discard any old and discolored leaves. Strip the leaves off the stems by grasping the base in one hand and pulling the leaves away from the stem with the other.

To clean the collards, fill the sink with cold water. Add the collards and stir vigorously with your hand; let the dirt fall to the bottom of the sink. Let the collards sit undisturbed for a minute or two. Carefully remove the collards from the water and place in a colander. Rinse out the sink and repeat the washing process 2 more times. After the third cleaning, carefully lift the collards out of the water, place in a salad spinner, and spin until dry.

Heat a large saucepan over medium-high heat. Add the fatback and cook for 5 to 10 minutes until it renders some fat. Add the onions, carrots, and ham hock and cook until the onion is a dark golden brown, about 25 to 30 minutes.

Add the greens to the pan and cook, stirring, until wilted. Add the broth, vinegar, Worcestershire sauce, soy sauce, thyme, and pepper. Cover the pot and simmer for 45 minutes, until the greens are tender.

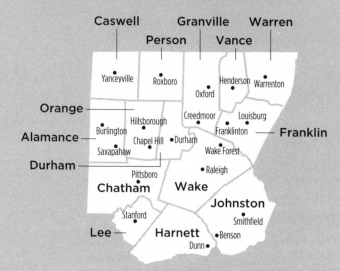

Caswell

Granville

Warren

Person

Vance

Yanceyville

Roxboro

Henderson

Warrenton

Oxford

Orange

Creedmoor

Louisburg

Hillsborough

Burlington

Franklinton

Franklin

Alamance

Chapel Hill

Durham

Saxapahaw

Wake Forest

Durham

Pittsboro

Raleigh

Chatham

Wake

Johnston

Stanford

Smithfield

Lee

Harnett

Benson

Dunn

Triangle

Like the Triad, the Triangle's fourteen-county region was heavily concentrated in tobacco, textiles, and furniture, with Durham once known as the tobacco capital of the world. In the 1960s, white-collar industries took off, thanks to the newly formed Research Triangle Park and surrounding universities. While many Triad farmers turned to grapes, many long-time Triangle farmers switched to produce. But perhaps the biggest change in the Triangle was a boom in sustainable farms as environmentalism and social activism arrived. The first wave came in the 1960s and 1970s, with another arriving over the past decade and still going strong. Residents responded by becoming local-food supporters, shopping at farmers' markets, joining CSAs, patronizing farm-to-table restaurants, and putting the Triangle on the national map as a culinary destination.

FARMS

Cane Creek Farm

"I've always had this animal thing and this sleep deprivation thing, and it merged in farming," said Eliza MacLean, who has drawn national attention for her work with Ossabaws, a rare breed of pigs from Ossabaw Island, off the Georgia coast. After managing hog herds for North Carolina A & T State University, Eliza, a single mom and former adventure racer, began her own Cane Creek Farm in 2003 in Saxapahaw, southeast of Burlington. Four years later, in search of more land, she joined forces with 600-acre Braeburn Farm in Snow Camp, which raises beef cattle. Eliza tends to her own twenty acres, and both farms' meat is now marketed under the Cane

Creek label. In 2009 Cane Creek opened a small on-farm store. On Saturdays Eliza is usually on hand to give tours, letting customers check out not only her pigs, but also chickens, ducks, sheep, goats, and assorted pets. "People get to see the connection to their food," she said, "and how Mother Nature hates a monoculture."

1203 Longest Acres Road, Snow Camp (Alamance County), 336-376-5620,
www.canecreekfarm.us. Farm store open Thursday and Saturday.
Tours during store hours and by appointment.

Herb Haven

Suki Roth turned her passion for plants, the earth, and alternative medicine into Herb Haven, an herb farm on six acres in Graham, west of Chapel Hill. Growing and selling medicinal and edible herbs is only part of what Suki does. She offers individual herbal consultations, gives classes on holistic healing, and has created a line of botanical teas, tinctures, oils, and salves under the name Suki's Blends. Long recognized as a mentor to area herbalists, Suki also conducts herb classes and guided walks through her garden. "I like to take folks around and introduce them to all the wild medicinal and edible plants we have here in the Piedmont. Teaching others about what's around us is really important to me."

7922 McBane Mill Road, Graham (Alamance County), 336-376-0727,
www.herbhaven.com. Classes and walks by appointment.

Baldwin Family Farms

From the time he was a boy, V. Mac Baldwin wanted to be a cattle farmer. His dream was realized in 1969, when he and his wife, Peggy, purchased two Charolais heifers. In 1981 they bought a 301-acre spread in Caswell County, eventually acquiring several hundred more acres and several hundred more of the French cows, known for their protein-rich lean meat from natural grazing. Along with their son, Craig, V. Mac and Peggy started direct-marking their Baldwin Charolais Beef in 2002, later adding a farm store, just south of Yanceyville. The Baldwins also sell their beef through their website, at farmers' markets, and through several Whole Foods stores. Their long-term goal is to build a dining facility with a kitchen for special events. We diners are waiting.

5341 Highway 86 South, Yanceyville (Caswell County), 336-694-4218,
800-896-4857, www.baldwinbeef.com. Farm store open
Monday through Saturday. Tours by appointment.

Little Meadows Farm

When Liese and Bob Sadler bought forty acres on a former tobacco farm in Caswell County, their soil was poor and the farmhouse had seen better days. Since they arrived in 2006, they have turned this peaceful spread set along a dead-end dirt road into a thriving "vegetarian farm" encircling a modest, well-maintained home. The adventurous couple, who spent 1998 to 2003 living on a sailboat, grow vegetables and raise laying hens, fiber sheep, and dairy goats. Liese taught herself to shear sheep, spin wool, weave, and knit, as well as make soap and cheese from the goats' milk. She can be seen spinning at the Rockingham County Farmers' Market and also teaches spinning, soap making, and farmstead cheese making. Little Meadows runs a brilliant "adopt-a-goat" and "adopt-a-sheep" program. For a nominal fee, sponsors get photos and updates from the farm, as well as wool or soap from the animals.

432 Wilson Road, Providence (Caswell County), 336-388-0295, www.littlemeadowsfarm.net. Sales and tours by appointment.

Sleepy Goat Farm

You won't hear many local farmers say, "I fell in love with goats in Iran," but that's what happened to Della Williams. She and her husband, Jon Dorman, shared a neurology practice in Danville, Virginia, and also spent a decade practicing medicine in the Middle East. They had bought a former 160-acre tobacco farm a few miles south, in North Carolina, as a weekend retreat. In 2003 they moved to the farm, taking on a few goats, which led to sixty. "Ethel was my first. She followed me everywhere," Della said. In 2004 the couple became licensed to make and sell cheese. A few years later they added rental lodging. "People who stay here like to pet the goats and play with the babies when we have them," said Della. For an extra fee, guests can spend a day learning to make cheese.

7215 Allison Road, Pelham (Caswell County), 336-388-0703, www.sleepygoatfarm.com. Monthly open-farm days May to August. Tours by appointment. Lodging $$

Ayrshire Farm

Bill Dow of Ayrshire Farm, near Pittsboro, became the state's first certified organic farmer in 1980. Three decades later, he's still farming without pesticides, though he dropped the certification. Bill also helped start the Piedmont Farm Tour and the Carrboro Farmers' Market, and is still active in both. On about three of his twenty-two acres, all held in a land-conservation trust, Bill grows a mix of vegetables, including greens, toma-toes, and peppers, as well as heirloom apples and blueberries. During the course of our thirty-minute visit in the late morning, three restaurant owners called, asking, "Hey Bill, what do you have today?" Bill welcomes visits from individuals and small groups because, he said, "It's terribly important to show people where their food comes from." He also helps groom future farmers. "Farming is a viable option for young folks — if they can afford the land."

602 Friendly Pooch Lane, Pittsboro (Chatham County), 919-542-5528.
Sales and tours by appointment.

Bluebird Hill Farm

Even when she's farming, Norma DeCamp Burns looks stylish. So it's not surprising that one of her passions is teaching homeowners how to blend production gardens into the landscape. "Some gardens can look tacky in the front of a house, when they could be a design element," she said. Her Bluebird Hill Farm is a case in point, where even the basic garden has decorative trellises, and other producing plants grow in nooks and cran-nies. Norma also tends to 500 lavender plants. The architect and former Raleigh city councilor moved to the thirteen-acre farm west of Raleigh in 1999 with her husband, Bob, now deceased. Norma gives garden tours to groups large and small, and sometimes includes a cooking demonstration. "Teaching is something I thrive on, especially about gardening. I feel like I can really open people's eyes."

421 Clarence Phillips Road, Bennett (Chatham County), 336-581-3916.
Sales and tours by appointment.

Central Carolina Community College Land Lab

Students at Central Carolina Community College in Pittsboro take care of a small farm that the school calls its Land Lab. It's an integral part of the school's sustainable agriculture associates degree program, the first such program in the country when it started in 2001. The tidy two acres, which the public can visit, contains a variety of organically grown vege-tables, herbs, and fruits, and a flock of laying hens. Students occasionally sell the produce and eggs at the local farmers' market, but most of it goes to a small CSA for students and staff.

764 West Street, Pittsboro (Chatham County), 919-542-6495,
1-800-682-8353, www.cccc.edu.

Mob Mentality

In late 2008, a group of landless farm fans and wannabe farmers in their twenties and thirties gathered to talk challenges and opportunities. Out of that meeting came the Crop Mob (www.cropmob.org), a collective of mostly young farmers building community by volunteering for work sessions at area farms.

The workdays are organized through e-mail and social media, leading Trace Ramsey, a Chatham County mobber, farmer, writer, and photographer, to describe the Crop Mob as "guerrilla agrarianism in the information age."

After being written about in national publications, the Crop Mob has inspired similar groups near and far.

Cohen Farm

Not long after graduating from college in Illinois in 1971 with a degree in plant science, Murray Cohen bought land outside of Pittsboro and made it into one of North Carolina's first and longest-running pesticide-free farms. Four decades later, he and his wife, Esta, raise cows, pigs, and chickens on their 220-acre Cohen Farm. In 2007 they had their hay and grain certified organic. "People come from all around for my hay," he said. "I have horse owners tell me, 'My horses ate everything but the string.'" The couple sells produce and meat at markets and at the farm and encourages families to visit. "We especially like to show children where their food comes from, and how we take care of the animals."

688 Van Thomas Road, Pittsboro (Chatham County), 919-742-4433, www.cohenfarm.com. Sales and tours by appointment.

Piedmont Biofarm

Circling the Piedmont Biofuels Cooperative in Pittsboro are three acres making up Piedmont Biofarm, run by Doug Jones. A lifelong sustainable farmer who moved to North Carolina in 1999 from upstate New York, Doug specializes in year-round production. "Using seasonal extension techniques from up north, I found I could farm year-round here," he said.

The farm sells produce through a CSA and at markets. Doug also is well respected as a plant breeder and seed saver. "Down the road I'd like to start some form of seed service," said Doug, who created a variety of Asian collards available through the seed company Fedco. He also works on variety trials for Seeds of Change and leads seed-saving workshops. Piedmont Biofuels runs weekly tours, which include information on the farm as well as on-site hydroponic and vermiculture operations.

220 Lorax Lane, Pittsboro (Chatham County), 919-321-8260,
www.biofuels.coop/food/biofarm. Piedmont Biofuels tours every Sunday.
Seed-saving and other workshops run through the Abundance Foundation,
919-533-5181, www.theabundancefoundation.org.

Duke Homestead State Historic Site

Certainly one of North Carolina's most historically important farms is the Duke Homestead State Historic Site in Durham, where the tobacco industry was born. It was here, in the mid-1880s, that Washington Duke started to cultivate tobacco after his cotton crop failed. He and his children then started a factory for processing smoking tobacco, an enterprise that became The American Tobacco Company, the world's largest tobacco company. At the restored homestead and adjacent tobacco museum, visitors learn about tobacco farming and trade, and the Duke family history. Guided tours visit the packhouse, curing barn, and family home. The farm grows about a quarter acre of tobacco yearly, which it uses during the popular Tobacco Harvest Festival in September. This event features costumed harvesters, who haul the leaf by mule to the curing barn, grade it, tie it onto sticks, cure it, and auction it off, all in one day.

2828 Duke Homestead Road, Durham (Durham County), 919-477-5498,
www.dukehomestead.nchistoricsites.org.

Elodie Farms

When Dave Artigues bought a local farmer's herd of dairy goats, cheese-making equipment, and even the recipes in 2003, he was set. Except that the Citadel-educated psychologist didn't really know what he was doing. He learned quickly. Elodie Farms now produces a wide array of top-rate cheeses and has opened up its twenty-one acres for farms tours, weddings, and other happenings. The most sought-after events here are Dinners on

Guest chef Sam Poley puts the final touches on the goat-cheese appetizer at Dinners on the Porch, hosted by Elodie Farms in Durham County. Photo by Selina Kok.

the Porch, where visiting chefs whip up gourmet meals in the farmhouse kitchen, with many dishes featuring Elodie cheese.

Elodie Farms, 9522 Hampton Road, Rougemont (Durham County), 919-479-4606, www.elodiefarms.com. Dinners on the Porch March to December. Sales and tours by appointment.

Ganyard Hill Farm

When Milton Ganyard went to college, he was so intent on never going back to work on his family's dairy farm in Georgia that, he said, "I didn't quit until I had a PhD." After several stressful years working in environmental research, testing the effects of pesticides, Milton had a change of heart. In 1994 he, along with his late wife, Karen, opened the charming Ganyard Hill Farm, choosing to focus on farm education and tourism. "Our concept was we want people to have fun and learn while doing." The agritourism farm lives up to that ideal in many ways, including opening up rows of cotton, soybeans, sorghum, and corn. Children and adults flock to the cotton, picking and grinning. But the biggest draw is the four-acre U-pick pumpkin patch. "They're hard to grow in this climate, but we didn't want to truck them in like most people do."

319 Sherron Road, Durham (Durham County), 919-596-8728, www.pumpkincountry.com. Open late September to November.

Prodigal Farm

The goats made them do it. When Dave Krabbe and Kathryn Spann fled New York City in 2007 for life on the farm in Durham, where Kathryn grew up, they used a few of the four-legged farm friends to clear their overgrown land. Now their Prodigal Farm is home to some sixty-five goats. In 2010, the couple completed a state-of-the-art milking parlor and cheese-making room, and they now plan to sell cheese at area markets. The scenic 97-acre spread also includes a 120-year-old farmhouse, tobacco barns, and other historic outbuildings.

4720 Bahama Road, Rougemont (Durham County), 919-477-5653, www.prodigalfarm.com. Occasional public tours and by appointment.

SEEDS

What was once a dumping ground in a low-income section of Durham has become a magical green space open to the public. SEEDS, which stands for South Eastern Efforts Developing Sustainable Spaces, includes a small park and several garden plots on its two-acre parcel in Northeast Durham. Started in 1994 to bring a community garden to an impoverished area, SEEDS six years later added the youth program called DIG (Durham Inner-city Gardeners), in which teenagers grow produce, herbs, and flowers to sell at the Durham Farmers' Market. Today, DIG is flourishing and continues to expand. Also on the grounds are a twenty-five-plot community garden, a medicinal herb garden, and an outdoor art gallery.

706 Gilbert Street, Durham (Durham County), 919-683-1197, www.seedsnc.org. Open Monday through Saturday.

Hill Ridge Farms

What started in 1968 as a wholesale and U-pick tomato and strawberry farm became in 1986 one of the state's first agritourism businesses. Since then, Hill Ridge Farms, just north of Wake Forest, has also become one of the state's most popular, drawing more than 50,000 visitors a year. On the fifty-five-acre recreational farm, owner John Hill over the years has added barnyard animals, hay rides, a fish-feed dock, a giant slide, gem panning, and even train rides. When we visited on a weekday in October, a school group was taking a hayride through the back fields and small groups of parents and young children were strolling the paved walking paths, pint-sized pumpkins in hand.

703 Tarboro Road, Youngsville (Franklin County), 919-556-1771, www.hillridgefarms.com. Open April to November.

Lynch Creek Farm

Bob Lynch and Kerry Carter left their historic home in downtown Philadelphia for North Carolina in 1997 hoping to find another historic house in the Triangle. They ended up in a fairly new home on a fifty-five-acre property south of Henderson, which they've put into a conservation easement. The couple maintains their love of the old with Kerry's home-based antique showroom and Bob's transformation of their century-old packhouse into a stylish meeting space and vacation cottage. Bob, a computer scientist and jack-of-all-trades, also has built barns, hoop houses, and raised beds for an ever growing garden and farm, which includes produce, herbs, beehives, chickens, and grass-fed cattle. He also plans to host dinners at the farm featuring his own produce and meat. "That's my real passion—cooking and entertaining," he said.

1973 Rocky Ford Road, Kittrell (Franklin County), 252-492-2600, www.lynchcreek.com. Sales and tours by appointment. Lodging $$

Turtle Mist Farm

In moving from the Washington, D.C., area to North Carolina, Bob and Ginger Sykes wanted some land to build a house on, maybe twenty or thirty acres. Three years later, in 2005, they were living on sixty acres, with sheep, cows, ducks, turkeys, and a vegetable garden. They sell meat and produce off the farm and at markets. Before starting Turtle Mist Farm, said Bob, a retired human resources director, "most of my gardening was a tomato plant on the porch," though his interest was sparked by childhood memories of his great-uncle's farm. Bob and Ginger's land, north of Wake Forest, had previously been used as an amusement park and then for horse boarding and rodeos. Bob hopes to use the leftover facilities from those enterprises to hold family, business, and community events. The Sykes welcome visitors to the farm for day trips or overnight stays in their guest-house, across the pasture from the bigger home they built for themselves. "We want children to come and have the farm experience, just like I did."

221 Suitt Road, Franklinton (Franklin County), 919-702-2039, 919-457-2942, www.turtlemistfarm.com. Sales and tours by appointment. Lodging $$

Vollmer Farm

Few in the state do agritourism at the level Vollmer Farm does. What's most admirable is that the Vollmers operate a working farm, having made the switch from tobacco to produce, while also attracting thousands of visitors a year to their "Back Forty" entertainment complex. In the spring and summer, certified organic strawberries and blueberries are ripe for the picking, while some farm produce, snacks, and ice cream are for sale in the large gift shop. Starting in mid-September, the action really picks up. Tractor rides take hundreds of visitors and school groups a day to the "back forty" acres, filled with games, playgrounds, mazes, and other agri-culturally themed attractions.

677 Highway 98 East, Bunn (Franklin County), 919-496-3076, www.vollmerfarm.com. Open April to October.

Triple B Farms

After fifteen years of raising hogs in industrial feedlots, Bailey Newton did an about-face and started to raise cows the natural way—grass-fed, free-range, and free of hormones and antibiotics. "I'd lost my wife to cancer, and I knew what I was doing wasn't a real healthy deal," said Bailey, who owns the eighty-acre Triple B Farms, fifteen miles north of Oxford near the

Pumpkins are plentiful during the fall at Vollmer Farm in Franklin County. Photo by Selina Kok.

Virginia line. He started his first calves in 2000 and since then has added pigs, turkeys, lamb, ducks, and chickens to the mix, all naturally raised. Bailey sells at a farmers' market and from the farm, which sits pretty along a sparsely traveled country road. He's happy to give tours when he has the time. "People are amazed when they come here how it doesn't smell. That's just a joy to me."

3564 Harry Davis Road, Bullock (Granville County), 919-693-4246.
Open Monday through Saturday.

Lazy O Farm

Tami Olive Thompson and her father, Jimmy Olive, started Lazy O Farm educational farm in 2002. "Daddy knew he couldn't keep farming, so we decided to do something for children," said Tami, a fifth-generation farmer near Smithfield. "Johnston is one of the fastest-growing counties, and the people moving here have no idea what a farm is. We wanted to teach the younger generation what a farm gives them, from clothing to food." On a small section of the 500-acre farm, which produced tobacco from 1830 to 2005 and still grows some produce, the Thompsons have set up homespun farm displays. The emphasis is on animals, including pigs, cows, sheep, turkey, goats, and rabbits. Inside one barn they've set up a fascinating natural-history museum of sorts, with bones, horns, eggs, and nests, all found on the farm.

3583 Packing Plant Road, Smithfield (Johnston County), 919-934-1132, www.localharvest.org. Tours and hayrides by appointment and during special events.

Tobacco Farm Life Museum

Just off Interstate 40 southwest of Raleigh lies a lovely testament to the state's tobacco-growing past. The Tobacco Farm Life Museum in Kenly not only contains exhibits on tobacco and the rural lifestyle from the late 1880s to the late 1930s but also houses the beautifully maintained Iredell Brown farmstead out back. When we stopped by, a group of local women were making "roses" out of tobacco leaves. "It's taken four people all day to make a dozen," said museum manager Elaine Richardson, who stocks the "flowers" in the museum's gift shop. The museum, open since 1983, puts on several events year-round. At Hog Killing Demonstration Day in January, visitors learn how to cut out meat and make sausage. The ham the group salts and cures is eaten during Fall Heritage Day in October.

709 Church Street, Kenly (Johnston County), 919-284-3431, www.tobaccofarmlifemuseum.org.

Gross Farms

Gross Farms' fifteen-acre corn maze, the largest in the state, gets most of the attention on this century farm an hour southwest of Raleigh, but the three-acre pumpkin patch is also a huge draw, said Tina Gross. "They can go out and pick their own pumpkins, which you can't find at many places." The long-time tobacco growers started the maze, pumpkins, and hayrides in 2002, two years after they opened their popular three-acre U-pick strawberry patch. The berry season also kicks off sales at their produce stand, housed in an old tobacco barn. From there, the farm sells its own vegetables as well as those from other regional farms.

1606 Pickett Road, Sanford (Lee County), 919-498-6727, www.grossfarms.com. Open April to August, maze and pumpkin patch September to November.

Anathoth Community Garden

This five-acre garden is a peaceful protest to a violent crime. In 2004 Bill King, a white man who was married to a black woman and ran a small general store near where Anathoth now stands, was murdered. Upset by that crime, as well as by poverty and drug use, black resident Valee Taylor paired with Grace Hackney, pastor at Cedar Grove United Methodist Church, attended by whites, to hold a prayer vigil. That bond led Taylor to donate family land for the garden. Community members work the land and share in the bounty of vegetables and flowers. Anathoth also runs a year-long internship program for local teens and garden workshops for adults. Visitors are welcome to tour the organic gardens and stroll along the walking path, which is bordered by native plants.

3005 Lonesome Road, Cedar Grove (Orange County), 919-732-8405, www.anathothgarden.org. Tours by appointment.

Avillion Farm

For shepherd Elaina Kenyon, it's all about the fiber. Since learning to spin wool into yarn during graduate school, she's been smitten with its sources. Off a gravel road in rural Orange County, this full-time scientist working at Research Triangle Park tends to small flocks of angora goats (the source of mohair) and Shetland sheep, and about thirty angora rabbits in separate waist-level cages. If you think basic bunnies are cute, check out the angoras, whose fluffy long hair you'll want to bury your nose in. They're typically shorn four times a year, and their fiber is said to be eight times warmer than wool, said Elaina, who grew up on a meat and dairy farm. She sells some of her livestock, along with felting fabric, roving (preyarn fiber), and mill-spun yarns.

4737 Shanklins Dead End Road, Efland (Orange County), 919-563-0439, www.avillionfarm.com. Sales and tours by appointment.

Captain John S. Pope Farm

Second cousins Tommy and Bob Pope run Captain John S. Pope Farm, which has been in the Pope family since 1852. Its namesake was a Civil War captain who was Tommy's great-great grandfather and Bob's great-grandfather. Tommy, who lives a mile away, is the farmer, while Bob, a former marketing executive with Dupont, is the city slicker who lives in Raleigh. Throughout the years, the Popes have raised chickens, pigs, beef and milk cows, and tobacco. They were looking to diversify after the tobacco buyout when Bob spotted a truck painted with sheep. He tracked down the owner and learned he was a breeder. "Two weeks later I bought thirty ewes and a ram," he said. Now they have about 350 to 400 grass-fed sheep running on seventy-five acres and sell the meat to top restaurants in the area and from the farm. Farm tours, by appointment, include a look inside the family's 1870 farmhouse, which is used for special functions. Occasionally farm dinners are offered as well.

6909 Efland–Cedar Grove Road, Cedar Grove (Orange County), 919-621-1150, www.dorperscedargrovenc.com. Sales and tours by appointment.

Chapel Hill Creamery

Driving the long dirt road to Chapel Hill Creamery, we passed a couple of dozen grazing Jersey cows, their low-hanging teats swaying as if they were marketing the farmstead cheese made here. Almost as soon as Portia McKnight and Florence (Flo) Hawley opened the creamery in 2001, they drew an avid base of fans, from consumers to chefs and retailers. Their list of cheeses includes a cow's milk feta, camembert varieties, and a tricky-to-make fresh mozzarella, all available at farmers' markets and grocers. The cows should be proud.

615 Chapel Hill Creamery Road, Chapel Hill (Orange County), 919-967-3757. Sales and tours by appointment.

Coon Rock Farm

One of the best-known sustainable farms in the Triangle is Coon Rock, largely because of the publicity it's gotten as the supplier and part-owner of the award-winning Zely & Ritz restaurant in Raleigh. (Also in the works is its own farm-to-table Eno Restaurant and Market in Durham, and the farm also supplies Piedmont Restaurant in Durham.) Farmer and software entrepreneur Richard Holcomb left his suburban Raleigh home in 2005 for an 1880s farmhouse on fifty-five acres in Hillsborough. In his diversified operation, he grows more than five acres of produce. He also pasture-raises chickens, sheep, goats, and pigs for meat, as well as hens for eggs. Except for the hum of traffic, you'd never guess Interstate 85 was just around the corner from the farm's gently rolling hills punctuated by Coon Rock, a huge granite outcropping and local landmark.

1021 Dimmocks Mill Road, Hillsborough (Orange County), 919-732-4168, www.coonrockfarm.com. Sales on Saturday mornings. Tours by appointment.

Fickle Creek Farm

Ben Bergmann and Noah Ranells started the seventy-four-acre Fickle Creek Farm from the ground up in 1999, having bought the land a few years earlier when they were doctoral students at North Carolina State University. Ben's background is in horticulture and forestry, while Noah studied animal and soil science, giving them a strong knowledge base for their diversified livestock and produce farm. "My first goal is to be a grower of really high-quality produce grown in a sustainable way, with minimal mechanical and no chemical input," says Ben. "All of the animals are here to do something." They keep small numbers of sheep, cattle, and pigs, and much larger numbers of chickens — about 1,000 laying hens and, in a year, some 1,800 broilers. They sell produce and meat at several farmers' markets. Everything about their farm is operated sustainably, including their passive solar home. The couple also operates a bed and breakfast, giving guests a taste of life on a working farm and some very delicious breakfasts.

4122 Buckhorn Road, Efland (Orange County), 919-304-6287, www.ficklecreekfarm.com. Sales and tours by appointment. Lodging $$

Garland Truffles

Franklin Garland is single-handedly responsible for bringing truffle growing to North Carolina. The electronics engineer and his wife, Betty, started Garland Truffles in 1992 on their thirty-five acres south of Hillsborough

Ben Bergmann places just-gathered eggs on an automated grader
at Fickle Creek Farm in Orange County. Photo by Selina Kok.

after first planting inoculated trees in 1978. A decade later, truffle-sniffing hounds finally located the buried treasure, growing on roots underground, and chefs came calling for the gourmet mushrooms. Since then the Garlands have led the charge for the movement, even coordinating a Tobacco Trust Fund grant program in 2004 that awarded truffle-ready trees to fifty farmers. Though the couple still grow truffles, they're now known more for their nursery, which sells inoculated filbert and oak trees. Their pricey farm tours are more like gourmet brunches in their art-filled home, with truffle appetizers and omelets, champagne, and, yes, facts about truffle farming. Franklin envisions setting up a truffle tasting room run much like a winery. Do we sniff a trend?

3020 Ode Turner Road, Hillsborough (Orange County), 919-732-3041, www.garlandtruffles.com. Sales and tours by appointment.

Maple View Agricultural Center

Allison Nichols's first job was scooping ice cream at Maple View Farm Country Store. Now she helps run the place. Spending time away at college and then as an elementary teacher for a year showed her that "kids can't make the connection between their food and a farm," said Allison. When Maple View founder Bob Nutter, a fifth-generation farmer who opened the Hillsborough dairy in 1963, gave her the opportunity to create something new there, she chose an agritourism and educational center. In 2009, Allison and "Farmer Bob" opened the remarkable Maple View Agricultural Center just up the street from the store. A 5,800-square-foot building holds function space and classrooms, which include exhibits on crops, soil, insects, and a hatching area for chicks and ducklings. Out back are barnyard animals and four acres of labeled gardens, one dedicated to traditional North Carolina crops. Allison may have left the school system, but she's still very much a teacher.

3501 Dairyland Road, Hillsborough (Orange County), 919-942-6122, www.mapleviewagcenter.com. Hours vary seasonally.

McKee's Cedar Creek Farm Cornfield Maze

While most corn-maze operators rely on an outside service to design and even cut their maze, David and Vicki McKee decided to do it themselves. In 2001 they started using computer-aided design software and surveying tools to plot out and prepare a two-acre kiddie maze and twelve-acre adult maze on their 190-acre farm, in the family since the late 1700s. "Vicki's the artist; she always designs it," said David, who turned to landscaping, beef cattle, and the fall maze after he stopped growing tobacco in 2000. For maze-going young ones, the McKees bring in some barnyard animals, while the older set is treated to a haunted maze on select evenings.

5011 Kiger Road, Rougemont (Orange County), 919-732-8065, www.mckeemaze.com. Open weekends late September to early November.

Niche Gardens

"Our plants aren't grown to look good at a certain date but in the way that's best for the plant," said Blair Durant, owner of Niche Gardens. The nursery and mail-order business, situated in a rural area west of Chapel Hill, has received national attention for its propagation and sales of native plants since opening in 1986. All plants are grown outside, naturally, on three acres surrounded by woods. The display garden also serves as a gallery for garden art created by regional artisans. On Saturdays during the spring and fall, staff members lead free guided nursery walks.

1111 Dawson Road, Chapel Hill (Orange County), 919-967-0078, www.nichegardens.com. Hours vary seasonally.

Pickards Mountain Eco-Institute

Planted at the base of a 350-acre tract twenty minutes west of downtown Chapel Hill is Pickards Mountain Eco-Institute, an educational farm and sustainability learning center. It opened in 2008, a project of Megan and Tim Toben. Tim, a local green-building developer and former high-tech entrepreneur, bought the acreage and later put almost half into a conservation easement. The institute's centerpiece is its tidy gardens, filled with heirloom vegetables and some traditional North Carolina crops for demonstration, including heritage cotton varieties. A small number of livestock, including goats, pigs, and chickens, are raised here, too. A hybrid wind and solar system provides electricity for the farm, biodiesel refinery, and the living area, which houses about a dozen staff members and interns. Most programs here are for students, from kindergarten through college, but some public events and workshops are held throughout the year. And once a month, there's a potluck.

8300 Pickards Meadow Road, Chapel Hill (Orange County), 919-619-5475, www.pickardsmountain.org. Tours by appointment.

Day camper Jack Tibbs with his new friend at Spence's Farm in Chapel Hill (Orange County). Photo by Diane Daniel.

Spence's Farm

A short drive north from downtown Chapel Hill is a farm teeming with gardens, animals, children, and community. Spence Dickinson calls his thirty-five acres a community educational farm. It's been open since 1984. "We're modeled after the old family farm, where everyone had a job. We do the same thing here through small groups and mentoring." Indeed, on the Tuesday morning we visited, a daycare group was touring the sustainable gardens and chickens, one girl and her on-farm mentor were at the horse stables, and a local resident was taking a stroll around the gardens, berry bushes, pond, and hiking trails. The office is inside a renovated 1805 farmhouse, a reminder of the area's rural past.

6407 Mill House Road, Chapel Hill (Orange County), 919-968-8581, www.spencesfarm.com. Open daily.

Vista Wood Bison Ranch

"We believe in wholesome eating," said Jeff Peloquin, "and buffalo is one of the healthiest meat choices." That and a lifelong fascination with the North American bison, led the building contractor and his wife, Linda, to start Vista Wood Bison Ranch in 2003, when they moved to their eighty-five-acre farm north of Hillsborough. "I thoroughly enjoy the animals. They're intelligent and interesting," said Jeff, who sells the meat off the farm and at the Hillsborough Farmers' Market. He hopes to double the ten- to twenty-head herd, which often can be seen in the pasture next to the farm's driveway. "So many people were asking to come out that we finally opened the farm to the public in 2007." The couple takes groups on hayrides, explaining the historical relationship between bison and Native Americans and offering hands-on activities using bison horns, bones, and hide. Their long-range plan is to phase out the contracting business and open an events center on the farm—we hope with a bison-burger stand.

1020 Vista Wood Drive, Hillsborough (Orange County), 919-732-7554, www.vistawoodbisonranch.com. Sales and tours by appointment.

Walters Unlimited at Carls-Beth Farm

Back in 1990 Roland Walters asked his father, Carl, if he wanted to try rotational grass feeding at the family's cattle farm. "He said he did not," Roland recalled with a laugh. Fifteen years later, after living in Michigan for a decade, Roland moved back to Orange County with his young family to take over the business, called Walters Unlimited at Carls-Beth Farm. Finally he had a chance to try out his method. He also added goats, chickens (for meat and eggs), pigs, and turkeys and started selling cuts of meat from the farm. About 140 acres of the 410-acre spread are used for grazing, and visitors can see rolling meadows on each side of the road, dotted with grazing livestock.

7119 High Rock Road, Efland (Orange County), 336-212-4296, www.waltersunlimited.com. Sales Friday and Saturday afternoons and by appointment.

W. C. Breeze Family Farm Extension and Research Center

The aptly named PLANT (People Learning Agriculture Now for Tomorrow) harvests farmers while farmers harvest leased land at the W. C. Breeze Family Farm Extension and Research Center, an incubator farm managed by Orange County and North Carolina State University. The 168-acre farm was donated by Orange County native Colonel William H. Breeze and

his family to support local agriculture. The program introduces apprentice farmers to the basic how-tos of small-scale, sustainable vegetable and
fruit production. The public can drive by anytime to see what's growing or
to attend various events and workshops at the farm. Breeze Farm's most
popular event is its Farm to Fork Picnic fund-raiser, showcasing more than
fifty farmers and chefs and helping to fund future farmers.

4909 Walnut Church Grove Road, Hurdle Mills (Orange County),
919-245-2330, www.orangecountyfarms.org.

Whitted Bowers Farm
The first conversation Cheri Whitted and Rob Bowers had was about how
each of them dreamed of growing biodynamic fruit. Now married, the
couple moved from California to Orange County, where Cheri's family
has farmed for seven generations, to start the first certified biodynamic
farm in the Carolinas. Customers lined up almost as soon the fifty-five-
acre Whitted Bowers Farm opened in 2008. "The reaction has been stu-
pendous," Rob said. "People are thrilled that there's a biodynamic farm."
Biodynamic farming, a type of organic agriculture, emphasizes soil health
and the position of the moon and planets. While Rob acknowledges that
"occasionally we get rolling of eyes," many people are tuned in to the approach. Offerings on their growing roster include blackberries and blue-

berries, figs, heirloom tomatoes and melons, apples, sweet potatoes, and the state's only U-pick biodynamic strawberries. "People tell us, 'I've never tasted a strawberry this good.'"

8707 Art Road, Cedar Grove (Orange County), 919-732-5132,
www.whittedbowersfarm.com. U-pick strawberries in May.
Sales and tours by appointment.

Woodcrest Farm

You never know what Allan and Chris Green will be up to at their Woodcrest Farm, from caning chairs and blacksmithing to canning and holding a farm market from the wooden shed near their late nineteenth-century farmhouse. On their thirteen acres northwest of Chapel Hill, they grow produce and raise cattle, goats, sheep, ducks, and chickens. Allan has a blacksmith forge he fires up when the mood strikes. Chris teaches homestead skills. The couple offers self-guided tours around the farm and hayrides if there are enough visitors.

5604 Dairyland Road, Hillsborough (Orange County), 919-933-5105,
www.woodcrestfarmnc.com. Market open spring to fall. Tours by appointment.

Sunset Ridge Buffalo Farm

The view at Sunset Ridge Buffalo Farm is majestic—260 acres of undulating farmland punctuated by grazing bison. Farmer Jack Pleasant became fascinated with bison after he saw them at age fourteen during a backpacking trip at a Boy Scout camp in New Mexico. After a career in home health care, his thoughts returned to farming on land near Roxboro that had been in his family for two centuries. "I was interested in nutrition, and that led me to buffalo," said Jack. The family moved to the farm in 1994, got its first buffalo seven years later, and now maintains a herd of about 100. The Pleasants sell the meat off the farm by appointment and at farmers' markets. They also ship it. Farm tours are available for groups, and Jack and his wife, Sandy, plan to offer occasional open-farm days. "People often drive by, particularly during holidays, when people bring their out-of-town relatives, who don't believe there are buffalo in Person County."

465 Yarborough Road, Roxboro (Person County), 336-599-1297,
www.sunsetridgebuffalo.com. Sales and tours by appointment.

Green Planet Farm

Enough people offered to help Daniel Whittaker work at Green Planet Farm that he ended up establishing farm volunteer times during growing season. "It's a lot of inside-the-Beltline people who either wanted to start a garden and haven't or don't have a green area to get out to and get dirty." The farm's harvest goes to Green Planet Catering, billed as the region's first eco-friendly catering business. Daniel founded the company in 2008 and co-owns it. On a small portion of the twenty-acre farm east of Raleigh, he and partners grow such crops as peppers, squash, eggplants, lettuce, and tomatoes. "We'd like to be an educational facility some day," Daniel said. "I'd like other people to come out and learn, like I did."

2949 South Smithfield Road, Knightdale (Wake County), 919-832-6767, www.greenplanetcatering.com. Tours by appointment.

Historic Oak View County Park

Oak View farm, at one time 930 acres and four miles outside the Raleigh city limits, now sits alongside the Beltline and is called Historic Oak View County Park. Take away the constant din of highway traffic and this twenty-seven-acre, beautifully restored and maintained park could be deep in the country. As Wake County's finest reminder of its agricultural past (and one of the nicest educational farms in the state), the park is a magnet for school groups and families, especially during its many special events and festivals. First, check out the exhibits and pick up a self-guided walking tour map at the attractive Farm History Center, then head out to view the main farmhouse (built in 1855), pecan grove, plank kitchen, garden, small goat herd, cemetery, and cotton museum. Next to the museum is a demonstration cotton field, which likely will be some visitors' first view of cotton blooms and bolls.

4028 Carya Drive, Raleigh (Wake County), 919-250-1013, www.wakegov.com. Open daily.

The Little Herb House

Like the thousands of plants she raises, Lisa Treadaway's herb business south of Raleigh keeps growing. In 2000 she began planting an herb garden that has since multiplied many times over and blossomed into The Little Herb House. Inside and outside of former horse stables, Lisa has developed a 10,000-square-foot herb garden, a production field, a greenhouse for potted herbs (which she propagates and sells), a gift shop, classroom space, and an herb library. Over the years she has added several open-house events and classes, and morning and afternoon teas. Her garden is laid out in eight pie-slice sections, each with a theme, including Kitchen Garden, Butterfly and Bee Garden, and Potpourri Garden. Most interesting was the Herbs of the Bible Garden, containing plants commonly found in the Bible.

5800 Holland Church Road, Raleigh (Wake County), 919-772-3543,
www.littleherbhouse.com. Store hours vary seasonally.
Tours and classes by appointment.

FARM STANDS AND U-PICKS

Iseley Farms

Only a few miles north of downtown Burlington is the 250-acre Iseley Farms, in the family since 1790. The Iseleys, long a fixture in area, continue to farm through daughter Jane, who in 1981 joined her father, Edward, in running the tobacco farm. In her other life, Jane is a celebrated architectural photographer, with many books to her name. In 1989 she added wholesale vegetables, and in 1998 she started retail sales at the on-farm produce market, which is housed in a restored packing barn. The store has grown to include goods from neighboring farms. Families flock here in the spring to pick from three acres of strawberries; in the fall they come for farm-raised pumpkins and hayrides. Tomatoes are a big summer draw, as is the ice cream made by Maple View Dairy using Iseley strawberries and honey. Iseley Farms fans need not worry about the encroaching development; Jane has placed the land in an agricultural easement, ensuring its legacy.

2960 Burch Bridge Road, Burlington (Alamance County), 336-584-3323,
www.iseleyfarms.com. Market open April to Halloween,
strawberries April to June, pumpkins and hayrides in October.

Herndon Hills Farm

The Streets at Southpoint, one of the state's largest shopping malls, has transformed southern Durham from a sleepy rural area into busy suburbia. The picturesque 120-acre Herndon Hills Farm would almost surely

Consumers Are Taking a Cotton to Them

Innovative TS Designs, a T-shirt designer and printer in Burlington, remade itself after the 1994 North American Free Trade Agreement sent buyers overseas in search of bargains. Following their passion for protecting the environment, CEO Tom Sineath and President Eric Henry started selling organic cotton T-shirts and transformed their printing process from toxic to green. But each year, they discovered that more and more of their organic cotton yarn was coming from overseas sources.

Given that North Carolina is the country's fifth-largest producer of cotton (though not organic), in 2008 TS Designs started Cotton of the Carolinas (www.cottonofthecarolinas. com), a "dirt-to-shirt" initiative to produce a high-quality T-shirt made in the Carolinas (mostly North). The first harvest had a transportation footprint of less than 750 miles with a fully transparent supply chain. Online, TS Designs shows customers what was done where and by whom at every step of the way, from farm, gin, spin, knit, and finish to cut, sew, print, and dye. Try that on for size.

be gone by now if not for the family's decision to place fifty-six acres into a conservation trust. "It's an old family farm, and we all grew up here," said Nancy Herndon, who oversees the U-pick operation, with five acres of blueberries and ten of muscadine grapes. The U-pick part of what had been strictly a cattle farm started in 1990. "It was a way to accommodate the growing suburban environment, to give the farm a future."

7110 Massey Chapel Road, Durham (Durham County), 919-544-3313.
Blueberries June to July, grapes August to September.

Lyon Farms

Fifth-generation farmer Mark Lyon was the first to leave tobacco behind for produce at the family's farm in Creedmoor, north of Raleigh. That was in 1980, and Lyon Farms, which he runs with his wife, Rose, has been growing steadily since then. He added U-pick strawberries early on, and Lyon now has eight U-pick acres of strawberries, along with two of blackberries and four of tomatoes. Also grown on the 140-acre farm, of which fifty are in production, is a wide variety of produce, including corn, cantaloupes, peaches, and asparagus, which the Lyons sell from their produce stand, at farmers' markets, and through a CSA.

1544 Munns Road, Creedmoor (Granville County), 919-528-3263, www.lyonfarms.com. Open April to August.

Smith's Nursery and Produce Farm

Myron Smith transformed his family's tobacco farm into a plant nursery in 1980. Sixteen years later, he and his wife, Sarah, started a three-acre U-pick strawberry operation on their meticulously maintained fifty-acre farm. "Things kept evolving from there," Sarah said. Now Smith's Nursery and Produce Farm, just off Interstate 40 a half-hour south of Raleigh, grows and sells squash, zucchini, green onions, cabbage, and cantaloupes. The Smiths expanded the U-pick options to include blueberries and blackberries. Not stopping with produce sales, Myron and Sarah opened the farm to agritourism activities, including school tours, fishing in their stocked pond, and bird-watching. Seasonal events include a pumpkin patch and hayrides in the fall and an Easter-egg hunt in the spring. The farm also is a certified National Wildlife Federation Wildlife Habitat.

443 Sanders Road, Benson (Johnston County), 919-934-1700, www.smiths-nursery.com. Nursery open daily, produce sales seasonal.

McAdams Farm

The sign in front of McAdams Farm sports a cow, a tobacco leaf, and a strawberry. The golden leaf is a nod to its past. The farm was established in 1885, when the first McAdams moved to this land from neighboring Alamance County. On its 800-acre farm, the family grew the usual tobacco cash crop, as well as produce and livestock to feed family members. It still raises beef cattle (and also lamb), but in 2000 tobacco was replaced with strawberries, produce, and cut flowers. Now McAdams is run by fifth-generation farmers Howard and Karen McAdams. Growing methods are "part conventional and part sustainable," Karen said. Strawberries are

U-pick and already picked, and an on-farm market store has been running "forever," Karen said.

100 Efland–Cedar Grove Road, Efland (Orange County), 919-732-7701, www.mcadamsfarm.com. Hours vary seasonally.

Hilltop Farms

When Fred and Virginia Miller moved from Greensboro to her family's former tobacco farm outside of Raleigh in 1990, Fred became the only current family member to "get my hands in the dirt," he said. Virginia, meanwhile, opened a horse-boarding business. In 2002 Fred left his longtime office job to take up the life of a full-time farmer. Two years later, his thirty-acre Hilltop Farms became Wake County's first and possibly still only certified organic farm. Fred sells his vegetables through a CSA, an organic wholesaler, and, on Saturdays, out of a former tobacco shed turned farm stand. Fred also has one of the state's few organic strawberry fields, offering an acre of U-pick or prepicked strawberries.

6612 Kennebec Road, Willow Springs (Wake County), 919-552-5541, www.hilltopfarms.org. Open Saturday afternoons April to November. Tours by appointment.

Deans Farm Market

Deans Farm Market, which has been selling farm-fresh produce in Wilson since 1965, stepped things up a notch when farmer James Sharp bought the business in 2002. The action starts each April, when twenty-one acres of strawberries are opened for U-pick or prepick. James also ships the fruit through his Fresh-Pik Produce, the wholesale business that he started as a teenager, before attending North Carolina State University. Most of the bins at the market are stocked with produce from James's 350 acres. The market also sells shelled beans, honey from the farm's hives, and Deans's popular cooked and seasoned collards and cooked mustard salad. During October, kids can go on hayrides, visit barnyard animals, and get lost in a cotton maze.

4231 Highway 42 West, Wilson (Wilson County), 252-237-0967, www.deansfarmmarket.com. Open April to December.

FARMERS' MARKETS

Saxapahaw Rivermill Farmers' Market

The Saxapahaw Rivermill Farmers' Market is unlike any other farmers' market in the state. It's held in the evenings, bands perform, and alcohol is allowed, so it's more like a farmers' party. Musicians and farmers are situated in a parking lot, with the concert audience sitting on a grassy hill above. All this merriment is held at and sponsored by Rivermill Village under the name "Saturdays in Saxapahaw." The summer event, which started in 2005, has helped to create a buzz at this former textile mill that has been renovated into apartments and townhomes.

1616 Jordan Drive, Saxapahaw (Alamance County), 336-376-3122,
www.rivermillvillage.com. Held Saturday evenings May to August.

Durham Farmers' Market

Though the Durham Farmers' Market opened in 1998, it found a second life in 2007, when it moved downhill from a tucked-away parking lot to a covered pavilion in a gentrifying downtown area. In the first year of the move, some vendors reported that their sales doubled. With the frontage road closed to street traffic, an entire block swarms with pedestrians, giving it the feel of a street fair. The market features more than fifty farm, food, and craft vendors from within seventy miles of Durham. Chef appearances are held almost weekly during the summer. A favorite event is the Chef Challenge, when three chefs cook dishes from market fare, with a focus on a secret ingredient unveiled that morning.

501 Foster Street, Durham (Durham County), 919-667-3099,
www.durhamfarmersmarket.com. Held Saturday mornings
year-round and Wednesday afternoons May to September.

Carrboro Farmers' Market

Markets around the state have attempted to emulate the Carrboro Farmers' Market, considered to be not only a Carolina gem, but one of the best in the country. The festive year-round market showcases more than seventy-five farm, food, and craft vendors all, impressively, from within a fifty-mile radius of Carrboro, just west of Chapel Hill. Among them are some of the state's first and leading sustainable farms, including Ayrshire, Maple Springs Garden, and Peregrine. Held on Carrboro Town Commons, the

Farmers' Markets: A Legislative Act

Most farmers' markets around the country are operated by cities or private groups, but North Carolina also has five state-run markets. The first opened in Raleigh in 1955, the most recent in Lumberton in 1999. In between, markets have been established in Colfax (near Greensboro), Asheville, and Charlotte. All are run by the state Department of Agriculture, but they are quite different, with individual managers largely running the show.

While Raleigh is the only location to routinely turn a profit, the Lumberton market is the only one not deemed a success, as the market was built in a location with few produce farmers and few customers.

All state markets except Lumberton's have a wholesale and retail area. The produce in the wholesale area can theoretically come from anywhere in the country. Produce from the retail area should be from North Carolina farmers, though reselling is allowed, which means a nonfarmer vendor can sell produce from several farmers. The state markets have not established a system to inform the public about the sources of its offerings. Unless customers demand it, that's unlikely to change.

market, in business since 1979, attracts throngs of shoppers every Saturday morning, as well as a good number on Wednesday afternoons in season. Special events include cooking demonstrations and tastings held by bakers, chefs, and farmers. The annual tomato tasting features more than sixty varieties of tomatoes (we'd never seen so many in one place), along with recipes and tomato-growing tips from master gardeners. If you're not among the lucky locals, this market is worth a road trip.

301 West Main Street, Carrboro (Orange County), 919-280-3326, www.carrborofarmersmarket.com. Held Saturday mornings year-round and Wednesday afternoons April to October.

Midtown Farmers' Market

Being extremely mall-phobic, we expected to dislike the Midtown Farmers' Market at North Hills, a retail, office, and dining area in Raleigh. Although the scene is a bit corporate, with vendors canopies supplied by developer Kane Realty Corporation and imprinted with "North Hills," the offerings are homegrown and the farmers are local. Seeing shoppers stop for green beans and tomatoes before heading into a department store is a comforting sight, as is the gathering of families listening to live music on the nearby "Commons" green, even if the grass is synthetic.

4321 Lassiter at North Hills Avenue, Raleigh (Wake County), 919-881-1146, www.northhillsraleigh.com. Held Saturday mornings April to November.

State Farmers Market

More than half a century old and visited by about 3.5 million people a year, the State Farmers Market in Raleigh is the granddaddy of all North Carolina markets. On a Saturday during the peak season, some 30,000 people come here to browse, shop, dine, and socialize. This first of five state markets opened in 1955 near downtown and moved to its current site on seventy-five acres off Interstate 40 in 1991. The Farmers Building, open daily, is where more than fifty vendors ply their wares. Some grow everything they sell, while others grow none of what they sell, but all produce sold in this area should come from North Carolina soil. In the wholesale trucker shed, which many restaurants and produce stands frequent, the inventory could be from outside the state. The indoor Market Shoppes, near the Farmers Building, features products from about fifteen livestock farms and cheese makers and has one of state's best selections of North Carolina wine. During growing seasons, special commodity days are filled with activities for children and adults, making the market feel like party central, with everyone invited.

1201 Agriculture Street, Raleigh (Wake County), 919-733-7417, www.agr.state.nc.us/markets. Open daily.

Wake Forest Farmers' Market

Since opening in 2004, the Wake Forest Farmers' Market, in Wake Forest's charming downtown, has become a vital part of the community. "We have people whose weeks are just not complete without coming to the market," said manager Terri Wilkinson, who sells her baked goods under the name The Flour Garden. "We were started by farmers and are run by farmers,

Diana May, co-owner of Jordan Lake Christmas Tree Farm in Chatham County, pulls a tree through a baler to ready it for transport. Photo by Selina Kok.

with our own nonprofit," she said. About twenty-five agriculture and craft vendors participate on a given day, all are from within seventy-five miles of Wake Forest, and most farm sustainably. Given the size of the market, it's impressive that it operates year-round, even if on a scaled-back schedule during the winter.

110 South White Street, Wake Forest (Wake County), 919-570-1372, www.wakeforestmarket.org. Held Saturday mornings April to November, with reduced hours December to March.

CHOOSE-AND-CUT CHRISTMAS TREES

Jordan Lake Christmas Tree Farm

When Byron and Diana May took over Jordan Lake Christmas Tree Farm in 1994, they were fortunate to inherit customers from the previous owners, who started selling trees in 1987. The Mays, who live in the farmhouse on the twenty-four-acre property, grow twelve acres of choose-and-cut trees. While they offer beautiful native species, such as Virginia pine, Leyland cypress, Eastern red cedar, and Carolina sapphire, "most people go right for the Frasers," Diana said of the precut Fraser firs brought in from the mountains. She and her helpers also make a large selection of wreaths and operate a small gift shop with Christmas ornaments and accessories. On weekends, the family farm sets up hayrides and sells cider, hot chocolate, and homemade cookies.

2170 Martha's Chapel Road, Apex (Chatham County), 919-362-6300, www.jordanlakechristmas.com. Open late November to December.

Back Achers Christmas Tree Farm

The folks behind Back Achers Christmas Tree Farm, on the southern edge of Raleigh, are too clever. First they came up with a witty name (their backs ache tending to the back acres). Then they became the only tree farm we've seen to market to the state's growing Hispanic population by erecting bilingual roadside signs: "Arboles de Navidad Baratos." And, finally, the owners, father and son Frank and Bruce Barick, implemented an inventive way to keep the deer from rubbing their antlers on their five acres of trees—they scatter hair clippings around the tree. "People ask me, 'What's that hair-looking stuff under the tree?'" Frank said. "'Hair,' I tell them." A retired biologist and state wildlife commissioner, Frank once worked to restore the deer herd, so the joke's on him. Father and son grow a variety of pines and cedars, as well as sell precut Fraser firs, and maintain a tree database dating to 1954.

4533 Inwood Road, Raleigh (Wake County), 919-821-2071, www.back-achers.com. Open late November to December.

Pop-n-Son Christmas Trees

Pop-n-Son Christmas Trees has stayed true to its name ever since William (Pop) and Bill Stanton first planted trees on their Garner farm in 1974. When William retired, Bill's son, Cameron, stepped in. Like many eastern farmers, the Stantons quickly discovered they couldn't satisfy customers without offering some precut Fraser firs along with their own twelve acres of native pines and cedars. But unlike other growers, they started their own Fraser farm, in mountainous Sparta, in 1978. "We wanted to be able to control the freshness," said Bill. Pop-n-Son also sells "balled-and-burlapped" Christmas trees for later planting. It's easy to spot the hilltop farm. Just look for the giant blow-up snowman.

2518 Benson Road, Garner (Wake County), 919-772-0467, www.popnson.com. Open late November to December.

VINEYARDS AND WINERIES

Benjamin Vineyards and Winery

Nancy and Andy Zeman opened Benjamin Vineyards and Winery in 2004, on ten acres near the Haw River between Burlington and Chapel Hill. The Zemans are unusual on several counts. They grow muscadine and vinifera grapes, most of their grapes are cultivated without pesticides, and they run a zero-waste facility, meaning they recycle, compost, or reuse everything. Benjamin also sponsors one of the most interesting winery events around—a canoe paddle down the Haw River followed by a dinner using local farm ingredients, with wine pairings, of course.

6516 Whitney Road, Graham (Alamance County), 336-376-1080, www.benjaminvineyards.com. U-pick grapes August to October.

The Winery at Iron Gate Farm

When Debbie and Gene Stikeleather bought "the old Lynch farm" in 2000, it still had tobacco growing in the fields. Debbie, a native of Caswell County, had tobacco-farming relatives and grew up working in their fields and those of neighbors. While the couple initially bought their sixty-acre farm northeast of Burlington in order to preserve it, they ended up planting grapes in 2001. Three years later they started making their own wine, and opened The Winery at Iron Gate Farm. Their labels, with such names as Brightleaf White, Pack House Red, and Flue Fire, pay homage to the farm's past. While the rural setting is simple—a small building holding the tasting room and a gift shop—the couple has turned the winery into a lively entertainment venue by building a large patio and stage, with a lovely view of the vineyards.

2540 Lynch Store Road, Mebane (Alamance County), 919-304-9463, www.irongatevineyards.com.

Hinnant Family Vineyards

We were lucky enough to visit Hinnant Family Vineyards on a warm, breezy September day. The aroma of ripe muscadine grapes enveloped us during a stroll through a small portion of the seventy-five-acre muscadine vineyard, the state's largest. The vineyard's U-pick operation, in operation since 1972, was in full swing, while the tasting room, off Interstate 95 near Selma, had a steady flow of sippers. At the winery, which father-and-son dental professionals Willard and Bob Hinnant opened in 2002, you can buy not only wine and grapes but also grape plants. In addition to growing about ten varieties of muscadine grapes, Hinnant also makes berry wines and some drier muscadine varieties.

826 Pine Level–Micro Road, Pine Level (Johnston County), 919-965-3350, www.hinnantvineyards.com. U-pick grapes August to October.

STORES

Company Shops Market
Since its first event in 2007, the Burlington co-op Company Shops Market has signed up more than 1,500 owners and hopes to open its downtown store, housed in an old A & P grocery, by 2011. The plan to start a co-op was hatched years earlier by a small group of civic and business leaders in Alamance County wanting to find ways for local citizens to purchase food from nearby farms. The store is being modeled after the instantly popular Chatham Marketplace, a co-op grocery in Pittsboro.

268 East Front Street, Burlington (Alamance County), www.companyshopsmarket.coop.

Saxapahaw General Store and Café
When Jeff Barney and Cameron Ratliff took over the Saxapahaw General Store in 2008, it was a basic "cigarettes-and-gas convenience store," Jeff said. The shop, housed in what used to be the dye room of an old textile mill, still sells gas and cigarettes (now including organic smokes). But the owners have added high-end fare, with an emphasis on artisanal and local products, including produce, meat, cheese, and wine. The shop is a part of Saxapahaw Rivermill, an upscale housing project thirty minutes west of Chapel Hill. Another of Jeff and Cameron's transformations was turning a pizza-and-dog café into a made-from-scratch kitchen that emphasizes local products. "We use all local livestock and produce wherever possible," Jeff said. The eventual plan is to open a full-scale restaurant. What isn't likely to change? Gas and cigarettes.

1735 Saxapahaw–Bethlehem Church Road, Saxapahaw (Alamance County), 336-376-5332, www.saxgenstore.com.

Chatham Marketplace

As soon as Chatham Marketplace opened in 2006, its café became a local gathering spot. The co-op just outside of downtown Pittsboro became the first grocery in farm-friendly Chatham County to feature locally grown and organic products. We were struck by the volume of local offerings, including flowers, many choices of area wines, a full case of Cane Creek Farm meat and several regional cheeses. The produce section contained many items from local farms, as did many lunch and dinner items in the café. At the checkout counter we noticed farm-themed photograph cards, all created by Debbie Roos, a Chatham County extension agent. In 2009 Debbie planted and began maintaining a pollinator garden near the store's entrance, enticing bees as well as shoppers and diners.

480 Hillsboro Street, Pittsboro (Chatham County), 919-542-2643, www.chathammarketplace.coop.

Maple View Farm Country Store

During summer weekends, you can expect to wait in line at Maple View Farm Country Store, between Chapel Hill and Hillsborough. That's part of the fun at this farmstead ice-cream shop in the country, where queuing customers discuss their favorite flavors. Maple View Farm, started by Bob Nutter, opened its shop in 2001 after four decades of selling its milk and dairy products to other stores (which it continues to do). Along with ice cream treats, you can pick up farm-fresh milk, butter, and even beef. Housed in a building that looks like a farmhouse, complete with rocking chairs on the porch overlooking rolling meadows, the store is a magnet for families, country lovers, and bicyclists, who frequent these less-traveled roads. The pastoral view should stay intact because the Nutter family put 100 of the farm's more than 300 acres into a conservation easement. The farm has also opened two urban shops, but the original remains a classic.

6900 Rocky Ridge Road, Hillsborough (Orange County), 919-960-5535, www.mapleviewfarm.com.

A Southern Season

While A Southern Season doesn't specialize in farm-grown goods, its sprawling specialty foods department carries several North Carolina products, including locally made cheeses and condiments. This iconic re-

Everything Old Is New Again

Farm history is passed along in creative ways by Siler City artisan Roger Dinger, who finds new uses for old barns through his business, Reinbarnation (www.reinbarnation. com). Roger makes beautifully handcrafted furnishings and home accessories from mostly oak and pine North Carolina barns, the majority of them coming from farms being sold off to developers. "When I first visit these abandoned buildings I take photographs, talk to the owner, explore the grounds," Roger said. "I get a real sense of how it was used. In some small way, I'm recycling history." Before he sells a piece, Roger adds a tag to each item stating what county and farm the wood came from.

tail store supports local farms in many other ways as well. Since 2008, it has sponsored the South Estes Farmers Market, organized by a group of Orange County farmers, which sets up on the store's grounds year-round. Its restaurant, the Weathervane, works year-round with local vendors, farms, and products for its creative, upscale offerings. And the store's popular cooking school includes many farm-related sessions. One class even takes students to the farmers' market to buy the ingredients and then prepare them.

University Mall, 201 South Estes Drive, Chapel Hill (Orange County), 919-929-7133, 877-929-7133, www.southernseason.com.

Weaver Street Market

Weaver Street Market is more than one of the best food cooperatives in the country; it's the beating heart of funky little Carrboro. The community market, which opened its doors in 1988 and now has more than 3,500 members, carries an amazing selection of organic and local farm-fresh products, from produce, meats, and cheeses to wine and beer. Its hot and cold food bars, bakery, and café are always bustling, but the real action happens on the front lawn, where locals and visitors congregate. In 2002

Weaver Street opened a second, smaller location in Chapel Hill, and in 2008 it opened another full-size store in Hillsborough. The Hillsborough location also houses the (non)corporate headquarters and "Food House," where dishes and bread are prepared for all three stores.

101 East Weaver Street, Carrboro (Orange County), 919-929-0010; 716 Market Street in Southern Village, Chapel Hill, 919-929-2009; 228 South Churton Street, Hillsborough, 919-245-5050; www.weaverstreetmarket.coop.

Whole Foods Market

While local products constitute a minority of Whole Foods' offerings, the Texas-based chain has done much to support local growers, producers, and harvests in their immediate communities. In the Triangle, where most of North Carolina's Whole Foods stores are located, you can find, for example, meat from Mae Farm in Louisburg, lilies from Sarah and Michael's Farm in Durham, cheese from Celebrity Dairy in Siler City, and much more. The chain also has supported several state groups, including area farmers' markets, the Center for Environmental Farming Systems, and the Carolina Farm Stewardship Association.

Whole Foods Markets are in Raleigh, Cary, Durham, Chapel Hill, and Winston-Salem; www.wholefoodsmarket.com.

DINING

Yancey House Restaurant

It took awhile for Lucindy Willis to stop being surprised whenever she learned that customers from surrounding cities were making the drive to Caswell County just to eat at her and her husband Michael's restaurant, the upscale and comfortable Yancey House. Opened in 2005, it's housed in the original Bartlett Yancey home, constructed around 1807 and gorgeously restored by the couple. Lucindy, an inventive chef, has a PhD in English and taught at several North Carolina colleges and universities before retiring to the kitchen. The couple lives on a nearby forty-acre farm with a large garden and also maintains a garden on the restaurant grounds. Lucindy goes to the Carrboro Farmers' Market weekly and works with a list of farmers closer by. "The interesting part of where we are is people show up on the back porch with whatever they're growing." As for diners making the trek there, "We're in the middle of nowhere and everywhere, and it's a pretty drive either way."

699 Highway 158 West, Yanceyville (Caswell County), 336-694-4225, www.yanceyvillage.com. $$–$$$

Angelina's Kitchen

It all started with Angelina Koulizakis's CSA membership with Piedmont Biofarm in Pittsboro. Getting involved in the local food scene led her to a job cooking at Chatham Marketplace, which inspired Angelina to open her own place in 2009. Angelina's Kitchen relies on almost all local ingredients, from produce to meat, for the chef's scrumptious Greek and Greek-inspired dishes. "I want to show people how we eat in a Greek village, where everything is fresh," she said.

23 Rectory Street, Pittsboro (Chatham County), 919-545-5505, www.angelinaskitchenonline.com. $–$$

Fearrington House Restaurant

In the pasture at Fearrington Village, a former dairy farm turned planned community near Chapel Hill, visitors can watch some fifty Belted Galloway cows meander about. They will not, however, find the beloved black-and-white Belties on the menu at the Fearrington House Restaurant. Other local farms, however, do supply the state's only AAA Five Diamond restaurant with beef, pork, poultry, produce, and cheese, while seafood is delivered daily from the North Carolina coast. Working with executive chef Colin Bedford and other kitchen staff, Fearrington gardeners maintain all-organic produce and herb gardens, both in the ground and inside four greenhouses on the premises. While the restaurant keeps its produce for its own use, area farmers set up shop on Tuesdays from spring to fall at the Fearrington Village farmers' market.

2000 Fearrington Village, Pittsboro (Chatham County), 919-542-2121, www.fearrington.com. $$$

Four Square Restaurant

One July, chef Shane Ingram created a new menu featuring tomatoes every night for a month. He and Four Square Restaurant co-owner and wife Elizabeth Woodhouse kicked off their tomato love-in with a four-course menu highlighting tomatoes from several farms, with appearances by the farmers. Ever since opening the stately, upscale Four Square in 1999, Shane has worked with farmers from surrounding counties, even cooking heritage turkeys for the restaurant's popular Thanksgiving meal. Shane and Elizabeth also tend to their own gardens, at their home in Pittsboro and at the restaurant, housed in a city landmark, the 1908 Bartlett Mangum House. On Saturdays, Shane can usually be found at the Durham Farmers' Market, where he shops regularly and occasionally gives cooking demonstrations and tastings.

2701 Chapel Hill Road, Durham (Durham County), 919-401-9877, www.foursquarerestaurant.com. $$-$$$

Fullsteam Brewery

First, beermeister Sean Wilson started Pop the Cap, a group that successfully lobbied for lifting the alcohol-by-volume cap in North Carolina in 2005. Then Sean went full steam into his next project, Fullsteam Brewery. Sean calls his operation a "plow to pint" brewery focusing on southern ingredients. He secured a downtown Durham site in 2009, and the following year he brought in Brooks Hamaker—a beer consultant, writer, and former Abita Brewing Company executive. Sean's concoctions have included such farm-fermented beverages as Mothervine, a local wheat and scuppernong white beer; and Sweet Potato, with the spuds added to the mash. Sean named his brewery tavern, which opened in 2010, Fullsteam R & D in honor of his "southern mad science" pursuits.

726 Rigsbee Avenue, Durham (Durham County), 620-464-9274, www.fullsteam.ag. $

LocoPops

The Triangle went loco for LocoPops as soon as the Mexican-style popsicles were available from a tiny Durham storefront in 2005. Bringing *paletas* here was the brainchild of Summer Bicknell, who found a mentor in Mexico to teach her the process of turning fresh fruits, herbs, and spices into frozen treats. Since then, Summer, former business partner Connie Semans, and the staff have developed a couple of hundred flavors and rotate about twenty-five at any given time, many seasonal and sourced locally. For instance, a fall menu included pumpkin spice and a maple butternut squash *paleta*, not your typical lickable flavors. Over the years,

LocoPops has grown like kudzu, sprouting several more stores in the area. Speaking of kudzu, it's not on the menu. Yet.

1600 Hillsborough Road, Durham (Durham County),
919-286-3500, www.ilovelocopops.com. $

Magnolia Grill

Magnolia Grill opened in 1986 in a squat brick building that was once a health-food store. While the location's offerings have changed, its emphasis on freshness has not. More than two decades later, Ben and Karen Barker's culinary stage, now considered more a national destination than a neighborhood bistro, continues to highlight fresh ingredients from near and far. By being the pioneers they are, the Barkers have done more than satisfy diners and support local farms and markets. Along the way, the couple has trained a long roster of cooks who in turn have gone on to whip up their own versions of seasonal cooking using local ingredients. While some have moved on to other regions of the country, many have stayed in North Carolina. For that, we are thankful.

1002 Ninth Street, Durham (Durham County), 919-286-3609,
www.magnoliagrill.net. $$$

Piedmont

Piedmont was among the first in a new wave of farm-sourced restaurants in Durham when it opened in 2006. Since then, the original owners and chefs, Andy Magowan and Drew Brown, have gone on to other locally minded culinary ventures, and Piedmont in 2010 was purchased by the Eno Hospitality Group. One of Eno's partners is Richard Holcomb, owner of Coon Rock Farm in Hillsborough, which now sources the restaurant. Heading up Piedmont's kitchen is chef Marco Shaw, a Washington, D.C., native who previously owned a nationally acclaimed farm-to-table restaurant in Portland, Oregon. At Piedmont, Marco focuses on American dishes inspired by the seasons and, of course, Coon Rock's harvest. When extra shopping is needed, Marco need only walk up the street to the Durham Farmers' Market.

401 Foster Street, Durham (Durham County), 919-683-1213,
www.piedmontrestaurant.com. $$

The Refectory Café

The Refectory Café at Duke Divinity School is the answer to your prayers for delicious, wholesome meals in a cafeteria setting. And, yes, it's open to the public. Duke's first "green café" opened in 2005. It focuses on ingredients from local farms, offers many vegetarian and vegan dishes, and stresses recycling, from its compostable disposables to the refinished 1920s oak tables and 1940s dinnerware, pulled out of a Duke storage basement. The walls, multiple shades of green, are appropriately adorned with framed photographs of farm scenes. While a Refectory outpost at Duke Law School is smaller and less atmospheric, the food is equally divine.

Duke Divinity School, Westbrook Building, 919-668-3498; Duke Law School, corner of Towerview and Science drives, 919-613-8552, Durham (Durham County), www.therefectorycafe.com. $–$$

Scratch Baking

Durham artisan baker Phoebe Lawless of Scratch Baking put a new spin on community supported agriculture. Instead of offering a CSA, she sells subscriptions to a CSP — community supported pie — and sells savory and sweet pies at the Durham Farmers' Market. Flavors change constantly, as Phoebe gets most of her ingredients from local farmers. "Being limited to what I find at the market forces me to be so much more creative," she said. "My cooking skills have improved because of it." To the delight of her fans, she opened a retail outlet in downtown Durham in 2010, featuring even more pie flavors, other pastries, side dishes, and coffee.

111 Orange Street, Durham (Durham County), 919-489-9431, www.piefantasy.com. $

Watts Grocery

Chef Amy Tornquist started using local ingredients when shopping for her company, Sage & Swift Gourmet Catering. But when she opened Watts Grocery in Durham in 2007, she really put the spotlight on local sourcing, painting the slogan "local ingredients by the forkful" on the side of Watts' brick building and listing farms on the menu. The Durham native features southern food with an urban twist, and her cooking has received accolades near and far. Amy named the restaurant after her old neighborhood store, and, like the original Watts Grocery, her restaurant has become a local fixture.

1116 Broad Street, Durham (Durham County), 919-416-5040, www.wattsgrocery.com. $$

Crook's Corner

Crook's Corner chef and owner Bill Smith has been working with local farmers since the early 1980s, even before he moved to Crook's, the venerable New Southern restaurant started by the late Bill Neal. "I go to the market every week," said Bill, "but the bulk of my produce comes from people I've been dealing with for many years, older farmers, mostly retired, who bring stuff to the back door." Indeed, subscribers to Crook's regular e-mail list will hear about the weekly arrivals of such southern specialties as wild persimmons, Jerusalem artichokes, fresh figs, and more. Bill also uses local cheeses and hams, but it's the octogenarian farmers and other neighbors who really keep his menu revolving.

610 West Franklin Street, Chapel Hill (Orange County),
919-929-7643, www.crookscorner.com. $$

Lantern

When Andrea Reusing opened Lantern in 2002, it was just another small restaurant on Franklin Street in Chapel Hill. A few years later, the Asian-influenced Lantern was shining brightly, trumpeted in local and national publications for its extreme local sourcing and inventive dishes. Andrea, its celebrated chef, is invited to cook at national food events and has been nominated for a James Beard award. All the while, she continued to follow her course of serving simple, authentic, and superbly cooked Asian food using seasonal and local ingredients. She helped popularize pork from Cane Creek Farm's Ossabaw hogs, which remain a staple on the menu, and became an advocate through her work with Slow Food Triangle. Her first cookbook, *Cooking in the Moment*, was published in 2011.

423 West Franklin Street, Chapel Hill (Orange County), 919-969-8846,
www.lanternrestaurant.com. $$-$$$

Margaret's Cantina

The morning we visited Bill Dow of Ayrshire Farm we eavesdropped on a call from Margaret Lundy of Margaret's Cantina in Chapel Hill. "What've you got today, Bill?" asked Margaret for maybe the 6,000th time in their lives. Margaret opened a takeout chicken stand in 1990 and a restaurant in 1993. From the start, she purchased organic produce from Bill and other farmers, and she hasn't let up, also buying local meat, eggs, and poultry. In her "South meets Southwest" cuisine, you will find farm-fresh sweet potatoes, greens, tomatoes, chiles, and chicken, and her weekly specials use

Nourishing the State's Food System

One of the country's leading research centers for the study of sustainable food systems is here in North Carolina. The Center for Environmental Farming Systems (www.cefs.ncsu.edu), established in 1994, works to develop and promote environmentally friendly food and farming systems, involve communities, and provide related jobs. The center is a partnership of North Carolina State University, North Carolina A & T State University, and the state Department of Agriculture and Consumer Services. Much of the center's work, including organic produce and beef production, is done at its 2,000-acre research farm at the state's Cherry Farm in Goldsboro. Public outreach includes a series of workshops and an annual lecture with a national leader in the local-food movement. One of the center's largest and most recent projects is a study and action plan for building a sustainable local-food economy in North Carolina to reduce the state's reliance on distant food-supply networks.

the best of the current harvest. Or just ask Bill what he's got coming out of the ground and you'll likely find it at Margaret's.

1129 Weaver Dairy Road, Chapel Hill (Orange County), 919-942-4745, margaretscantina.com. $–$$

Neal's Deli

A taste of the big city came to small-town Carrboro in 2009, when Matt and Sheila Neal opened Neal's Deli. Their credentials for running a fresh-ingredient sandwich shop that relies on local sourcing are impeccable. Matt is the son of the late Bill Neal, who founded Crook's Corner and put the "new" in southern cooking, while Sheila used to run the Carrboro Farmers' Market, arguably the state's top producer market. Together they've created a breakfast and lunch mecca whose reputation quickly grew larger than its square footage. The menu features such bargain delicacies as sausage biscuits using local heritage pork and organic flour and sandwiches piled with homemade corned beef, pastrami, and sauerkraut. Is it lunchtime yet?

100-C East Main Street, Carrboro (Orange County), 919-967-2185, www.nealsdeli.com. $

Panciuto

The low-key Aaron Vandemark minimally markets his restaurant's commitment to working with area farmers. Yet Panciuto, the cozy Italian restaurant he opened in 2006, sits securely in the top five of the Triangle's farm-to-table eateries. Once diners are seated inside this elegant, comfortable Italian restaurant in downtown Hillsborough they'll see evidence of Aaron's relationship to the farming community. A small card on each table lists more than a dozen farms and artisanal food makers he does business with regularly. Flip it over and you'll find a long list of vegetables and meat products featured in the current menu "all available at the Durham and Carrboro Farmers' Markets." While Aaron may be telling us that we can gather these wonderful, fresh ingredients as easily as he can, most of us probably wouldn't end up making equally amazing dishes.

110 South Churton Street, Hillsborough (Orange County), 919-732-7261, www.panciuto.com. $$–$$$

Panzanella

Many food co-ops have cafés, but few have full-service restaurants like Panzanella in Carrboro. Opened in 2001 as part of Weaver Street Market, around the corner from the co-op, Panzanella focuses on organically and sustainably grown produce and livestock. The Italian-themed menu changes often to reflect growing seasons. The setting in the Historic Carr Mill Mall is notable as well, with exposed brick walls, hardwood floors, and high ceilings. Panzanella occasionally holds dinners highlighting the bounty of a local farm and schedules other events throughout the year, including the Local Farms/Local Art Exhibit in conjunction with the Piedmont Farm Tour.

200 North Greensboro Street, Carrboro (Orange County), 919-929-6626, www.panzanella.com. $$

Herons

Since the Umstead Hotel and Spa opened in Cary in 2007 as the area's first luxury hotel, its restaurant, Herons, has been evolving into a more farm-fed operation. Chef Scott Crawford, who became executive chef in 2009, quickly joined in at farm-to-fork events and even established a large garden on the Umstead grounds. Herons already had a fifteen-bed herb garden on site, maintained by the chefs, and a commitment of two acres at Blue Sky Farms in nearby Wendell, the restaurant's sole produce supplier. Goat cheese comes from Elodie Farms, and beef and pork from Cane Creek Farm. Future plans at the hotel and restaurant include garden tours and farm-fresh cooking classes.

100 Woodland Drive, Cary (Wake County), 919-447-4200, www.heronsrestaurant.com. $$–$$$

Poole's Downtown Diner

It's a good bet that a chef who learned to cook from Lantern's Andrea Reusing is going to seek out local farmers and food artisans. Ashley Christensen, an already well-known chef in Raleigh, opened her own restaurant, Poole's Downtown Diner, in 2007 to much deserved fanfare. Inside the landmark 1940s building, which once housed a pie shop and then a luncheonette, Ashley has maintained the old-fashioned diner decor but given the menu a thoroughly modern freshness. (One caveat: it's quite noisy.) Her seasonal, locally focused menu items, written bistro style on chalkboards above the double-horseshoe bar, change weekly if not daily. As Ashley noted, "We work with local growers and artisan producers to showcase their craft, while practicing our own."

426 South McDowell Street, Raleigh (Wake County), 919-832-4477, www.poolesdowntowndiner.com. $$

Zely & Ritz

When chef Sarig Agasi teamed up with Coon Rock Farm in 2004 to supply produce and meat for Zely & Ritz, they brought the first farm-to-table restaurant to Raleigh, and one of the few then in the Triangle. Sarig grew up on a kibbutz, an Israeli agricultural community, and understood the importance of fresh, naturally grown ingredients. The cuisine is Mediterranean and Middle Eastern, much of it in small servings, or tapas. All locally sourced items on the menu are printed in green, which makes it the prevailing color.

301 Glenwood Avenue, Raleigh (Wake County), 919-828-0018, www.zelyandritz.com. $$–$$$

The resident goats welcome guests to The Inn at Celebrity Dairy in Chatham County. Photo by Selina Kok.

LODGING

The Inn at Celebrity Dairy

By the time you roll out of bed at The Inn at Celebrity Dairy, fifty miles west of Raleigh, farmer and innkeeper Brit Pfann will have been up for hours, feeding and milking dozens of goats and tending to hens. You have the option of joining him or waiting for the evening rounds. After farm chores, Brit makes guests' breakfast, which includes fresh eggs and goat cheese made by his wife, Fleming. The bed and breakfast opened in 1997, a decade after the Pfanns started the 330-acre Celebrity Dairy, whose cheese is sold at farmers' markets and in stores. Rooms are in a Greek Revival farmhouse, the sitting area is the original settler's 1800 log cabin, and a two-story atrium joining the buildings serves as the dining room. The public is invited here for monthly Sunday dinners and occasional open-barn days.

144 Celebrity Dairy Way, Siler City (Chatham County),
919-742-5176, www.celebritydairy.com. $$

SPECIAL EVENTS AND ACTIVITIES

Franklin County Farm Tour

Several regions in the state host farm tours, but it's unusual for a county to sponsor its own on a regular basis. The Franklin County Farm Tour, which started in 2004, is organized by the county extension service, namely, Martha Mobley, a farmer herself. She and her husband, Steve, produce pasture-based Angus beef at their Meadow Lane Farm in Louisburg, one of about a dozen tour stops.

Franklin (Franklin County), 919-496-3344,
www.franklincountyfarmfresh.com. Held in May.

Coats Farmer's Day

In 2012 the small town of Coats will celebrate a century of Farmer's Day celebrations, making it one of the oldest farmers' days in the Southeast. The still-rural area of about 2,100 residents, just west of Benson, was created by farmer James T. Coats, who bought 700 acres in 1875 and later opened a general store on his property. Coats Farmer's Day draws about

20,000 visitors and includes a track show, watermelon races, a pig cook-off, and a street dance.

Downtown Coats (Harnett County), 910-897-6213,
www.coatschamber.com. Held in October.

Touchstone Energy North Carolina Cotton Festival

You know the annual cotton festival in Dunn has come of age when it becomes the Touchstone Energy North Carolina Cotton Festival and upward of 10,000 attend. Dunn, just off Interstate 95 near Interstate 40, was once surrounded by woodland and cotton fields and had an active cotton trading business. The festival, started in 1998, pays homage to the town's agricultural past. Activities, held on eleven blocks in downtown Dunn, include a classic car show, crafts, food, and music; and visitors can take a shuttle to a cotton gin nearby that's fired up for the weekend.

Downtown Dunn (Harnett County), 910-892-3282,
www.nccottonfestival.com. Held in November.

Benson Mule Days

Since its start in 1949, Benson Mule Days has grown from one day to four, and it brings about 65,000 people to this town of 3,500, near the intersection of interstates 40 and 95. Friday is dedicated solely to the animals, with mule jumping, pulling, and racing events, while Saturday's highlight is one of the state's largest parades, with more than 2,000 horse, mule, and buggy entries. Throughout the event, there's a carnival, games, live music, loads of food, and even a daily rodeo.

Downtown Benson (Johnston County), 919-894-3825,
www.bensonmuledays.com. Held in September.

Got to Be NC Festival

More than 1,000 pieces of antique farm equipment, a fiber fair with spinners and weavers, a carnival, a draft-horse pull, and lawn mower races keep customers busy at the Got to Be NC Festival, held in May at the state fairgrounds in Raleigh. Our favorite part of the weekend event, which the Department of Agriculture started in 2008, is the Food and Wine Expo, showcasing products from dozens of North Carolina food companies and wineries. In addition to wine, items for sampling include peanuts, salsa, and sauces. The expo also has a demonstration stage featuring area chefs and cookbook writers, as well as a barbecue cook-off.

1025 Blue Ridge Road, Raleigh (Wake County), www.ncagfest.com. Held in May.

Hen-side the Beltline Tour d'Coop

City slickers flocking to a tour of backyard chicken coops sounds like something you'd hear at a comedy club, but it's no joke. Hen-side the Beltline Tour d'Coop, a benefit for Urban Ministries of Wake County, has grown each year since it was started in 2006 by the capital's hen-happy residents. One Saturday a year, some twenty backyard farmers in Five Points and other neighborhoods open up their coops to the curious. Yards range from small to tiny, and the chickens come in all varieties, from run of the mill to prize winners. If you're left wanting more, ask cofounder Bob Davis about his Chicken Keeping 101 course.

Raleigh (Wake County), www.hensidethebeltline.blogspot.com. Held in May.

North Carolina State Fair

With some 878,000 attendees over ten days, the North Carolina State Fair, which debuted in 1853, is the largest agricultural event the state puts on. Amid the deep-fried delicacies, games of chance, and altitude-altering rides are hundreds of farm-friendly events. The State Fair Ark showcases more than sixty animals, including lambs, goats, cows, turkeys, and dozens of varieties of chickens. In the Exposition Center you can see giant pumpkins and watermelons, prize-winning vegetables of all sorts, and honeybees. There and in other locations, animals and their owners vie for prizes during the daily Livestock and Poultry Competitions. A stocked tobacco barn is open to the public at Heritage Circle, where during the week simulations show the locally grown tobacco being cured and taken to auction.

1025 Blue Ridge Road, Raleigh (Wake County), 919-821-7400, www.ncstatefair.org. Held in October.

Tomatopalooza

What started in 2003 as a backyard party has turned into a major tomato-tasting event, most recently held at Lake Wheeler Park in Raleigh. At Tomatopalooza, hosted by Lee Newman and Craig "Tomato Man" LeHoullier, some 200 varieties of mostly heirloom tomatoes are tasted and voted on. Craig, who lives in Raleigh, is famous in tomato and foodie circles for rediscovering and then preserving the Cherokee Purple breed. Gardeners are encouraged to bring their ripe 'maters to share.

Raleigh (Wake County), www.tomatopalooza.com. Held in July.

Piedmont Farm Tour and Eastern Triangle Farm Tour
Like the plants you'll see in this yearly spring event, the Piedmont Farm Tour keeps growing and growing. Co-sponsored by Carolina Farm Stewardship Association (CFSA) and Weaver Street Market in Carrboro, the tour started in 1995 with fewer than a dozen farms. Now about forty dot the self-guided route, and some 3,000 families attend the April event. The CFSA bills the weekend as "the nation's largest farm tour," a credible claim. In 2006 the CFSA added the Eastern Triangle Farm Tour in September, giving Triangle residents even more visiting options. That self-guided tour has grown to include about twenty stops, from urban to suburban and rural farms. Both tours include a mix of sustainable produce farms, livestock, nurseries, vineyards, and educational agriculture projects. While some of the farms on the tour are always open to the public, this is a chance to view others that typically aren't.

www.carolinafarmstewards.org, 919-542-2402. Piedmont tour held in April, Eastern Triangle tour held in September.

Taste Carolina and Triangle Food Tour
A sure sign that the Triangle area is becoming a food destination is the arrival of not one but two food-tour outfitters. Triangle Food Tour, based in Raleigh, came on the scene in 2007, a side project of food lovers Leigh and Peter Eckle, who run an executive relocation company. They started with the Raleigh Walking Food Tour and expanded into other towns. Taste Carolina, which offered its first tour in 2008, is run by Durham food-obsessed business partners Lesley Stracks-Mullem and Joe Philipose. With both groups you'll be treated to tastings of entrées, desserts, and beer and wine. During the restaurant stops, chefs impart behind-the-scenes tidbits, including how they work with local farmers. For its Chapel Hill tour, Taste Carolina includes a visit to several vendors at the Carrboro Farmers' Market.

www.tastecarolina.net, 919-237-2254; and www.trianglefoodtour.com, 919-434-8978.

Let's Hand It to Him

If you go on an organized farm tour or visit certain farms and festivals throughout the state, chances are you'll see a contraption called "use-yer-foot" (www.useyerfoot.com). The portable hand- and dish-washing stations are the brainchild of Kevin Meehan of Turtle Run Farm in Alamance County.

Foot pedals release soapy and clean water, allowing hand washing in places without conventional plumbing. They're lightweight, foldable, and can be rented or bought.

"I ship them all over the country," said Kevin, who came up with the idea during a weeklong music festival near Robbinsville in 1987. He named it after the instructions for operating this clever homegrown device: just use yer foot.

RECIPES

Goat Cheese Scones

These scones created by Inn at Celebrity Dairy innkeeper Brit Pfann in Chatham County highlight the farm's freshly made goat cheese. Nonguests shouldn't feel left out; the dairy also sells the pastries at local farmers' markets.

MAKES 12 SCONES

2	cups all-purpose flour
¾	cup whole wheat flour
3	tablespoons sugar
1 ¼	teaspoon baking powder
1	teaspoon baking soda
½	teaspoon salt
8	tablespoons butter (1 stick) chilled, cut into ½-inch cubes
1	cup crumbled goat cheese (about 7 ounces)
½	cup raisins
1	cup well-shaken buttermilk

Preheat the oven to 425 degrees.

In a large bowl combine the flours, sugar, baking powder, baking soda, and salt. Using a pastry blender or two knives, cut the butter and goat cheese into the dry ingredients until it resembles coarse meal and some pieces of butter are the size of small peas. Carefully mix in the raisins. Add the buttermilk and stir slightly (do not overmix) until the dough comes together in moist, large clumps.

Turn the dough onto a lightly floured surface. Carefully bring together and knead 3 or 4 times. Divide the dough in half and pat each half into a 6-inch round. With a large knife, cut the rounds into 6 even wedges. Place on a baking sheet about 1 inch apart and place on the middle rack of the oven. Bake until golden brown, about 20 minutes. Transfer to a rack to cool slightly. Serve warm or at room temperature with your favorite preserves.

Vegan Cauliflower and Walnut Soup

Refectory Café Chef Danielle Mitchell substituted coconut milk for skim milk to make this lip-smacking soup vegan. It's a lunchtime favorite at the café's two Duke University locations, which are open to students, staff, and the lucky public.

SERVES 4

1	medium cauliflower, about 2 pounds, coarsely chopped (most pieces should measure ¾ inch, some should be larger)
1	medium yellow onion roughly chopped, about 1 ½ cups
2	cloves garlic, minced
3 ½	cups vegetable broth/stock
1	14-ounce can coconut milk, well mixed
5	tablespoons walnut pieces, toasted and coarsely chopped
1	teaspoon dried oregano
1	teaspoon dried basil
½	teaspoon dried tarragon
	Kosher salt and freshly ground black pepper
	Pinch of paprika
1	tablespoon Italian flat-leaf parsley, chopped

Place the cauliflower, onions, garlic, and vegetable broth in a large pot. Bring to a boil, lower the heat to a simmer, cover, and cook until tender, about 5 minutes. Add the coconut milk, 3 tablespoons of the walnut pieces, and the oregano, basil, and tarragon.

Pour half the soup into a food processor or blender (you may have to do this in batches) and puree until smooth. Add back to the remaining soup. Adjust the spices if necessary and season with salt and pepper. Reheat the soup and serve with a sprinkle of paprika, chopped parsley, and the remaining walnuts.

Pink-Eye Purple Hull Peas

Raleigh cookbook writer and Edible Piedmont *publisher Fred Thompson is wild about southern pink-eye purple hull peas, or shell beans. "They don't turn brown, which gives them better plate appeal, they stay nice and firm, and make a great cold pea salad. Just add other diced veggies and a vinaigrette." This recipe is from his upcoming* Southern Sides *cookbook.*

SERVES 8

2	strips applewood smoked bacon, cut into a small dice
½	cup yellow onion, peeled and cut into a small dice
2	ribs of celery broken in half
3	cloves peeled garlic, whole
1	bay leaf
2	sprigs of fresh thyme tied with a string
4	cups water
4	cups shelled pink-eyes or other shell beans (about ⅔ of a bushel in the shell)
2	tablespoons balsamic vinegar
	Salt and freshly ground black pepper

Place the bacon in a 5-quart pot over medium heat. Slowly render the bacon until the bits are crisp and brown. Remove with a slotted spoon, place on a paper towel, and reserve. Remove all the fat but a tablespoon or so. Add the onions and sauté until wilted and soft, about 10 minutes. Stir in the celery, garlic, bay leaf, thyme, and water. Simmer this mixture for 20 minutes. Remove the celery and garlic.

Add the peas and cook for 20 to 25 minutes (do not overcook). Drain the peas, reserving the cooking liquid. Toss the peas with the balsamic vinegar. If serving immediately, add salt and pepper to taste and a little of the cooking liquid. If using as a component in a recipe or serving later, cool the liquid and then add it to the peas and refrigerate. Reheat very gently in the cooking liquid to avoid mushy peas.

From *Southern Sides Cookbook: Mouthwatering Dishes That Really Make the Plate*, by Fred Thompson. Copyright © 2011 by Fred Thompson. Used by permission of the University of North Carolina Press. www.uncpress.unc.edu.

Crab Salad with Lemon Mayonnaise

"If you're lucky enough to find fresh-picked crab meat, this is a nice spring or summer salad," said Aaron Vandemark, chef and owner of Panciuto in Hillsborough. At the restaurant, he has served it with fried-green tomatoes or black beluga lentils with olive oil.

SERVES 4 AS AN APPETIZER

FOR THE CRAB

1	pound cleaned lump or jumbo lump crab meat
1	tablespoon finely minced jalapeño
3	scallions, thinly sliced (about ⅓ cup)
2	tablespoons extra virgin olive oil

FOR THE MAYONNAISE

	Zest of ½ lemon plus 2 tablespoons lemon juice
1 ½	teaspoon apple cider vinegar
1	clove garlic, roughly chopped
1	egg yolk
¾	teaspoon dry mustard
½	teaspoon hot sauce
1	tablespoon good quality mayonnaise
2	tablespoons water
¼	cup melted butter, cooled
½	cup canola oil
	Pinch cayenne pepper
6	basil leaves, coarsely chopped
	Salt and freshly ground black pepper
¼	cup olive oil

In a large bowl combine the crabmeat, jalapeño, and scallions and mix lightly.

In the bowl of a food processor or blender, combine the lemon zest and juice, cider vinegar, garlic, egg yolk, dry mustard, hot sauce, mayonnaise, and water. Combine the melted butter and canola oil in a liquid measuring cup. With the food processor running, add the oil mixture, first in small drips, then in a slow steady stream until the mixture has been emulsified. Add the cayenne, basil, salt, and pepper and pulse until combined.

Add at least half the lemon mayonnaise and olive oil to the crabmeat and lightly fold together with a rubber spatula (do not break up the crab meat). Add more mayonnaise and olive oil if desired.

Northampton

Pasquotank
Perquimans · Camden

Gates · Currituck

Hertford

Halifax · Halifax

Bertie · Elizabeth City · Jarvisburg

Hertford

Edenton · Chowan

Edgecombe

Rocky Mount

Nash

Williamston · Washington

Martin · Tyrrell · Dare

Manteo

Wilson

Wilson · Pitt

Greene · Greenville · Washington · Hyde

Montgomery

Goldsboro

Troy

Moore

Cumberland

Snow Hill

Wayne

Beaufort

Kinston · Craven

Lenoir · New Bern · Pamlico

Southern Pines

Hoke

Sampson

Trenton · Jones

Rockingham

Fayetteville

Clinton

Kenansville

Richlands

Laurinburg

Red Springs

Richmond

Lumberton · Dublin

Scotland

Bladen

Burgaw

Morehead City

Onslow

Carteret

Duplin

Pender

Robeson

Columbus

Tabor City

Wilmington

Brunswick

Shallotte

New Hanover

Coastal Region and Sandhills

The coastal region and the Sandhills, a swath of ancient beach dunes that divides the Piedmont from the coastal plain, are the state's most rural section. North Carolina's first exports came from here—tar, pitch, and turpentine created from the plentiful pine trees. These days, sweet potatoes, berries, peanuts, and more thrive in the sandy soil. Meanwhile, industrial animal agriculture, mostly hogs, turkeys, and chickens, has replaced traditional row crops, such as cotton, corn, and tobacco, as the leading source of income. Most of the farms in this forty-two-county region use conventional growing methods, with only a handful certified organic. The wine made in these parts comes from native muscadine grapes. At Duplin Winery in Duplin County, the state's highest-volume standalone winery, some 100,000 guests a year slide up to its long tasting counter for a homegrown sweet treat.

FARMS

Shelton Herb Farm

Margaret Shelton doesn't have just "basil," at Shelton Herb Farm; she has more than two dozen types of the herb. Overall, you'll find more than 600 varieties of herbs at the business she started in 1986 on her family's 200-year-old home site just west of Wilmington. Shelton, who propagates everything she sells, grows culinary and medicinal herbs, dozens of lettuces, and several varieties of flowers. A few turkeys, pigs, and ducks also share the two acres, which stay open to the public year-round. Not content to rest on her bay laurels, Shelton has started to graft citrus trees.

340 Goodman Road, Leland (Brunswick County), 910-253-5964, www.localharvest.org. Open year-round.

A Day at the Farm

Sisters Julia Bircher and Melissa Barnett grew up on the last remaining dairy farm in Craven County, run by their father, Woodrow McCoy. The McCoy Dairy closed in 1993, after a fifty-six-year run. But thanks to the sisters, much of the farm's history has been lovingly preserved in an agritourism operation called A Day at the Farm. Public events are held spring to fall, and many groups, from corporate teams to Scouts, hold outings here. Activities include hayrides, butter churning, and tours of the two dairy barns, milking parlor, smokehouse, corn crib, and a kitchen filled with vintage utensils. For outdoor lovers, there's a pond, a nature trail, a playground, and gardens. An ice cream stand is open during warm months. The loft in the packhouse is perfect for sleepovers, where friendly ghosts of farm seasons past are sure to visit.

183 Woodrow McCoy Road, Cove City (Craven County), 252-514-9494, 252-514-3033, www.adayatthefarm.com. Open March to November. Tours by appointment.

Gillis Hill Farm

You know there must be pressure on Gillis Hill Farm owners to sell to developers when the Fayetteville Wal-Mart is only a mile away. But instead, the Gillises turned to agritourism to help save the 300-acre family farm, and the results are quite impressive. (The family also grows row crops on 2,000 acres in neighboring counties.) We were lucky to have eighth-generation Andrew Gillis, born in 1980, as our tour guide the day we stopped by. He led us through the 1900 farmhouse, which is now used as an office, meeting area, and ice cream shop. Family photos decorate the rooms, and an introductory film stars then kindergartener Andrew giving a farm tour. Outside, a well-marked walking tour wanders past a smokehouse, chicken coop, tobacco barn, sawmill, pond, and newly built waterpowered grist mill. A greenhouse was transformed into an educational center and auditorium. If you time it right, you can find a kilt-clad Andrew honoring his Scottish heritage by playing the bagpipes.

2701 Gillis Hill Road, Fayetteville (Cumberland County), 910-867-2350, www.gillishillfarm.com.

Island Farm

Throughout the decades, when Roanoke Island went from being farmland to the tourist stopover it is today, the two-story mid-1800s farmhouse of the Etheridge family managed to survive several changes of ownership. In the 1980s then Manteo Mayor John Wilson and three cousins, all Etheridge descendants, donated the house and a half acre of land to the Outer Banks Conservationists. After a decade of painstaking and costly restoration work, the nonprofit group opened the home place as Island Farm in 2010. On the eleven acres (more land was later acquired), reconstructed period outbuildings detail the farming life, including a corncrib, barn, dairy, smokehouse, and slave quarters. Farm demonstrations bring to life this important piece of Roanoke Island's history at its only remaining pre–Civil War farmstead.

1140 Highway 64, Manteo (Dare County), 252-473-6500, www.theislandfarm.com. Open spring to fall.

A. J. Bullard's Orchard

"I do have a lot of interests," agreed A. J. Bullard, retired dentist, former semipro baseball player, musician, amateur herpetologist, and self-taught horticulturist. What he's most known for these days is A. J. Bullard's Orchard, which he started in 1967 and has put on the map of horticulture groups across the state if not the country. On five acres south of Mount Olive, A. J. grows about forty species of fruit and nut trees containing at least 150 varieties of fruits, nuts, and grapes, making his the largest collection of fruit and nut trees in North Carolina. Sadly, for us, A. J. doesn't sell his harvest. Species include various berries, exotic fruit trees, pomegranates, figs, pears, pecans, English and black walnuts, and more than thirty kinds of muscadine grapes. "It's just something that keeps growing."

264 Farrior Road, Mount Olive (Duplin County), 919-658-4424. Tours by appointment.

Tarkil Branch Farm's Homestead Museum

On this 300-acre traditional corn and soybean farm, Benny Fountain pays homage to his family and farming. With Tarkil Branch Farm's Homestead Museum, he and his wife, Annette, have preserved their past in twelve restored buildings and 850 artifacts. While the farm started in 1912, the Fountain family has lived in the area since the 1800s. "I wanted to show how folks used to live," Benny said. The treasure trove, which opened to

Where Are All the Peanut Trees?

While Georgia is the nation's top peanut producer, North Carolina comes in fifth, with an annual output of about 359 million pounds, most coming from eastern counties. So where are all the peanut trees? Unlike nuts, such as walnuts and pecans, peanuts are legumes and grow in the ground. Their oval-leaf plants reach about eighteen inches tall, while the "pegs" grow down into the soil and are dug up during harvest, like potatoes.

the public in 2003, features a 1830s "dogtrot" farmhouse (two buildings connected by a breezeway), a tobacco barn, corn crib, and smokehouse. The old chicken coop has been transformed into a museum with household goods, tools, school supplies, and even everyday items such as hair curlers and household bills. The homemade display signs make the place all the more endearing. The final stop on the tour is a 1925 country store that had been in business a mile down the road. "When the owner's son called me and said, 'How would you like the store?' I almost passed out," Benny said. "That's exactly the kind of thing I want hear."

1198 Fountaintown Road, Beulaville (Duplin County), 910-298-3804, 910-296-4235, www.tarkilfarmsmuseum.com. Open Saturdays and by appointment.

Rainbow Meadow Farms

If you've had local poultry, pork, or lamb at a North Carolina restaurant, there's a good chance some of it was raised on Rainbow Meadow Farms in Snow Hill. The multigenerational family farm sells at markets and to retail outlets as well. In operation since 1746, it started selling naturally raised livestock in 1996. For years, farmer Jeff Garner grew commercial chickens for Perdue farms. One day Jeff tasted a chicken dish made from the free-range flock of his daughter, Genell Pridgen. "I said, 'Honey, what did you do with this chicken to make it taste so good?' And she said, 'Daddy, it's because it's raised out on the pasture.'" he said. Now father and daughter, along with their spouses, Sandra and John, raise chickens, hogs, lambs, and cattle. Their lambs are Dorper, chickens are from the French stock La Belle Rouge, pork is Berkshire and beef is American Devon. "That's what all the chefs at the upscale restaurants told us they wanted," Jeff said. Tastes good to us.

1065 Lloyd Harrison Road, Snow Hill (Greene County), 252-747-5000, www.rmfpasturepuremeats.com. Sales and tours by appointment.

John L. Council Farms

After living in New Jersey for thirty-five years, John Council returned home in 1994 to take care of his father. Since then, John L. Council Farms, run by John and his extended family twenty-five miles west of Fayetteville, has established a reputation for its sustainable farming methods. They were honored for those practices in 2009 after winning the Gilmer L. and Clara Y. Dudley Small Farmer of the Year award from the Cooperative Extension Program at North Carolina A & T State University. On the family's sixty-eight acres in Hoke County, a former tobacco farm, John and crew raise five acres of produce, heritage breeds of hogs, chickens, turkeys, cows, and some goats and rabbits. They sell from the farm, at markets, and to area restaurants. "Farming is all I've wanted to do," John said. "I love seeing things grow."

1920 Haire Road, Shannon (Hoke County), 910-303-1546.
Sales and tours by appointment.

Raft Swamp Farms

Jackie and Louie Hough set out to buy twenty acres of land and ended up with ten times more. The retired soldiers, who met at Fort Bragg after their last tours of duty, grew up on or around farms in the Midwest. A desire for a little land led them to a deal they couldn't refuse, some 200 acres southwest of Fayetteville in Hoke County, which they called Raft Swamp Farms. They kept fifty for themselves, and turned the rest into a nonprofit farm incubator, a place for new farmers to lease land and learn to farm organically and sustainably. They also placed the land, which abuts Raft Swamp, on the North Carolina Birding Trail, and birders have wasted no time in wandering these lowlands, spotting dozens of species. The couple grows and sells berries, peaches, asparagus, lavender, shiitake mushrooms, and tends to their own beehives. Hedgerows for wildlife habitat mark off incubator plots, with walking trails connecting them. A picturesque windmill, which pumps well water for irrigation, forms a lovely exclamation point over the flat landscape.

4978 Red Springs Road, Red Springs (Hoke County), 910-977-0950, www.raftswampfarms.org. Sales and tours by appointment.

Crystal Pines Alpaca Farm

The menagerie at Crystal Pines Alpaca Farm, situated on twenty-five acres along a country road near Carthage, will draw your attention. First, there are the twenty or so alpacas, not your everyday Sandhills site. When we stopped by in June they had recently been sheared, with only little tufts left near the end of their legs. The herd is protected by Great Pyrenees livestock guard dogs, which are larger than some of the other livestock here — Pygmy goats, miniature donkeys, miniature horses, and even a miniature cow. They share the farm with peacocks, swans, ducks, chickens, and a pot-bellied pig. Owners Joe and Ursula Picariello, who retired here and opened the farm in 1999, built a large barn to house a farm store. From there, they sell plants they cultivate, eggs from their chickens, and hand-crafted goods made from their alpacas' fleece.

200 Holly Ridge Road, Carthage (Moore County), 910-947-6649, www.crystalpinesfarm.com. Sales and tours by appointment.

Malcolm Blue Farm and Museum

Not far from downtown Pinehurst and flanked by homes sits the quaint Malcolm Blue Farm, a cluster of historic buildings set on seven acres. Its namesake, Malcolm Blue, settled and farmed the land in the early nineteenth century. During the Civil War, Union soldiers took over the house, which is part of the National Civil War Trails. The farmstead, on the National Register of Historic Places, features several structures, including the packhouse, corn crib, gristmill, and the original home, filled with early 1800s furnishings. Of the buildings, only the farmhouse is open, and only for guided tours. Three popular events take place here yearly: a summer bluegrass festival, a Christmas open house, and, biggest of all, the three-day Historical Crafts and Farmskills Festival, which has featured craft and livestock demonstrations for more than four decades.

1177 Bethesda Road, Aberdeen (Moore County), 910-944-7558, www.malcolmbluefarm.com.

Finch Blueberry Nursery

While Finch Blueberry Nursery in Bailey sells plants, not the berries themselves, we admit to enjoying a little snack while wandering through the thousands of plants on this 250-acre farm, one of the world's largest and oldest suppliers of blueberries. The nursery, east of Raleigh, was started in the late 1940s by the late Jack Finch and was taken over by his son, Dan, in 1976. "If you ever bought a blueberry plant at Lowe's, more than likely it started here," said Dan, a well-regarded potter, who has a studio and gallery here as well. He also sells eastern bluebird houses through the nonprofit Homes for Bluebirds, started by his father in 1973 to restore eastern bluebird populations. Thanks in large part to his efforts, the bluebirds are back.

5526 Finch Nursery Lane, Bailey (Nash County), 252-235-4664, www.danfinch.com.

Fisher Pumpkin Farm

Linda Fisher has been dealing in pumpkins since she was a child. "I used to sell them on the side of the road," she said. "I only missed one year, and that was for college." Linda is a full-time cattle farmer, having taken over the family business outside of Rocky Mount from her father. She also spent several years as a high school and middle school teacher, where she learned, to her dismay, that children don't know much about farm life. "People's granddaddies and grandmas aren't on the farm anymore," she said. So in the 1980s Linda started to turn a large roadside pasture into Fisher Pumpkin Farm every fall, where she shows children and families about traditional farm life, including wool spinning, butter making, and cow milking. Animals are displayed alongside the harvest from her twelve acres of pumpkins, gourds, Indian corn, and fall squash. In 2009 Linda won a statewide environmental stewardship award for protecting the farm's natural resources.

4713 Red Oak Boulevard, Rocky Mount (Nash County), 252-443-4439.
Open in October.

Mike's Farm, Country Store, and Restaurant

From its humble beginnings in 1986 as a Christmas tree farm, Mike and Theresa Lowe have built a small agritourism adventureland in this Onslow County community known as Back Swamp. The sixty-four acres of family land, once a tobacco farm and now known as Mike's Farm, Country Store, and Restaurant, is home to a strawberry field, pumpkin patch reached by hayride, petting zoo, holiday light show, bakery, large gift shop, and, since 2001, a 150-seat country-style restaurant. With piped-in music and a shop full of imported tchotchkes, the farm has drifted some from its agricultural roots, but the almost constant crowds don't seem to mind. As for the Christmas trees, they're still here, but in far fewer numbers. "We sold 1,400 at the height of it, before stores like Lowe's starting selling them," Mike said. "Now we sell about 200. Nowadays, space for parking is more important than space for the trees."

1600 Haw Branch Road, Beulaville (Onslow County), 910-324-3422, 888-820-3276, www.mikesfarm.com.

Whispering Dove Goat Ranch and Apiary

In 1998 Linda and Dale Klose decided to leave Jacksonville proper and buy thirteen acres in the country. They'd talked about getting a couple of goats to clear the land, but nothing came of it until Linda was detoured onto a back road and spotted a goat farm. Within two months, in 2001, they had their first small herd. The couple, now full-time farmers at their Whispering Dove Goat Ranch and Apiary, sell goat meat and use the milk from their forty or so goats to make soap and skin products, all of which they sell at farmers' markets. "We started having people ask us for lamb, so now we have sheep, too," Linda said. And rabbits, chickens, guineas, turkeys, and a few pot-bellied pigs as pets. Later the couple added beehives, and keep fifteen to twenty, whose honey the Kloses sell or use in products. Linda and Dale lead tours on their tidy and well-organized farm and also work as consultants for would-be goat owners.

689 Harris Creek Road, Jacksonville (Onslow County), 910-455-7123, www.localharvest.org. Sales and tours by appointment.

John Blue House and Cotton Gin

Thanks to a host of volunteers in Laurinburg, farm buildings, an antique cotton gin, and a spectacular historic home have been restored and are open to the public. The John Blue House and Cotton Gin, set on twelve acres, tells the story of life and farming in the late nineteenth and early twentieth centuries. The house is a stunning piece of architecture from 1898 built in the image of a Mississippi riverboat. Its owner, John Blue, was credited with inventing an early fertilizer spreader; he also repaired cotton gins and made farm implements that were known worldwide. Also

African American Farmers United

Gary Grant, a long-time activist in Halifax County and president of the national Black Farmers and Agriculturalists Association (BFAA; www.bfaa-us.org), works with groups around the country "to help farmers stay in farming—if that's what they want." Gary's passion was fueled in part by his own family's struggles. "My parents' farm was foreclosed on here in Halifax county, along with nine other black families in 1978." When family members, himself included, tried to take over the debt, the Farmers Home Administration (now the Farm Service Agency) said the full balance would have to be paid, he said. "That's just the way things were."

Since 1920, black farmland has declined by half, to about 7 million acres, while white farm ownership has remained steady. "We're down to something like 1,400 black farmers in North Carolina," Gary said.

The BFAA was founded in 1997, when farmers joined for the class-action lawsuit *Pigford vs. Glickman*, which proved discriminatory practices. A decade later, with nearly $1 billion paid out to settle some of the discrimination claims (in lots of $50,000), some farmers are still wrangling with the government.

One of the group's big events is its annual Land Loss Summit. "Landowners, farmers, researchers, and land activists come together to talk about the situation and look at solutions," said Gary.

Gary and the BFAA are based in Tillery, another hallmark of black history. The once thriving community, ten miles northwest of Scotland Neck, was one of eight U.S. resettlement farms for black homesteaders created in 1935 as part of President Franklin Roosevelt's New Deal. The resettlement was liquidated in 1943, with fewer than 100 families buying their farms, of which most were later lost. The Remembering Tillery Museum (www.cct78.org) tells their stories.

on the grounds are a working cotton gin (fired up by mule power at the yearly John Blue Cotton Festival) and several historic buildings collected from the region, including a 1920s stocked country store, a furnished log cabin from the early 1800s, and a reconstructed hay baler.

13040 X-Way Road, Laurinburg (Scotland County), 910-276-2495, www.johnbluecottonfestival.com. Open year-round. Festival held in October.

Governor Charles B. Aycock Birthplace State Historic Site

Is it wrong to wish that a state historic farm site would be turned into a bed and breakfast? The Governor Charles B. Aycock Birthplace is a beautiful 1870s farmstead set on thirteen acres northwest of Goldsboro. When we poked around in the spring near closing time, life was quiet on the farm, save for the chickens, rooster, and a few bleating sheep. The view of pastures and woods conjured images of relaxing on the porch with a glass of sweet tea. Things can get a bit noisier here when school groups and families tour the grounds and buildings, which include an 1893 one-room school, an 1870 house, and a detached kitchen. On special days, costumed interpreters demonstrate farm chores, lye soap making, and open-hearth cooking. The staff maintains a small demonstration field garden with corn, cotton, and wheat; and a kitchen garden with heirloom vegetables—the perfect place to pick ingredients for our breakfast.

264 Governor Aycock Road, Fremont (Wayne County), 919-242-5581, www.aycockbirthplace.nchistoricsites.org.

FARM STANDS AND U-PICKS

Pee Dee Orchards

Since 1961 Pee Dee Orchards has been a favorite stop among locals and beachgoers from Charlotte and Greensboro traveling along Highway 74. The farm stand, run by Chesley Greene and two of his children, stays focused on the right things—fresh peaches and fresh peach ice cream (as well as a few other flavors). More than a dozen peach varieties picked from the families 150 acres of trees keep the fruit coming throughout the summer and beyond.

11279 Highway 74 East, Lilesville (Anson County), 704-848-4801. Open May to October.

Southside Farms

One of the state's most impressive farm stands and U-picks, especially considering its out-of-the-way location, is Southside Farms outside of Chocowinity in Beaufort County. In 1998 Shawn and Tracey Harding, using former tobacco land in Shawn's family, turned a few acres into a beautifully maintained and inviting shopping and picking area. Tidy rows of U-pick veggies line the drive to the greenhouses. Inside the greenhouses are tomatoes and flowers for sale. Other potted plants, herbs, and flowers are sold at the store, housed in a former tobacco packhouse. We went during strawberry season—Southside has two acres of U-pick strawberries and an acre of blackberries—and the place was hopping.

320 Harding Lane, Chocowinity (Beaufort County), 252-946-2487, www.southsidefarms.com. Open April to August.

Terra Ceia Farms

Drive toward Terra Ceia Farms in Pantego, thirty minutes east of Washington, North Carolina, and you could be in the Netherlands. For miles you see nothing but windswept, low-lying farmland, interspersed with canals here and there. The only thing missing are the windmills. It was here that a community of Dutch immigrants started farming in the mid-1900s. Terra Ceia, one of the largest bulb and cut flower suppliers in the country, is now run by the three sons of one of those immigrants, Leendert Van Staalduinen, who founded the farm and died in 2009. We arrived in late May, just after a major harvest, so we missed the blooms. But the sales area, a nondescript room where hundreds of labeled shelves are filled with bulbs, was open. (Most of the farm's business is by mail order.) Terra Ceia grows flowers on 1,200 acres, about 250 for cut flowers (Whole Foods is a major customer), mostly peonies and sunflowers, and the rest for bulbs, said co-owner Carl Van Staalduinen. In the bulb category, you'll find a little of everything, from allium to zantedeschia.

3810 Terra Ceia Road, Pantego (Beaufort County), 800-858-2852, www.terraceiafarms.com. Open September to January and March to June.

Holden Brothers Farm Market

When you eye Holden Brothers Farm Market's vast variety of vegetables, you'll find it hard to believe that this 200-year-old farm not all that long ago produced only tobacco, soybeans, and corn. Inside the bustling roadside store, shoppers can easily spot Holden's helpings amid other fresh and processed offerings thanks to signs on the old-fashioned wooden bins reading "Grown on Our Farms." Because the market sits along a highway near the Atlantic Ocean, vacationers return each year. Holden Brothers also has U-pick strawberries and tomatoes.

5600 Ocean Highway West, Shallotte (Brunswick County), 910-579-4500.
Open March to December.

Indigo Farms

Sixth-generation farmer Sam Bellamy and several family members offer visitors much to see at their combination certified organic farm, retail store, bakery, and nursery. The farm, behind the store, is known for its sweet and juicy U-pick strawberries, but also grows blackberries, blueberries, and vegetables. Livestock raised here includes fowl, sheep, turkeys, and the famed "NASPig" racers, which run on Farm Heritage Day, the first Saturday in October. In October and November, the Bellamys set up hay rides and a hay bale maze. Unlike a corn maze, it's dark, so you have to feel your way through, with no baling out.

1590 Hickman Road, Calabash (Brunswick County), 910-287-6794, 843-399-6902,
www.indigofarmsmarket.com. Open year-round. Tours by appointment.

Garner Farms

With its location right off Highway 70 just west of Morehead City and Atlantic Beach, the Garner Farms produce stand is one popular place, especially on weekends. The 100-acre farm is spread out in different spots around Newport. "I can be at any of them in five to ten minutes on a tractor, and half of that is due to traffic," said third-generation farmer Clayton Garner. The action starts mid-April, when customers descend on the farm's one acre of U-pick strawberries. In the sizeable open-air store, bins are filled with produce, most of it from the farm. Also sold at the stand are jams, jellies, salsas, and even pastries and pies, made by Clayton's sister, Sheila Garner, in their Garden Patch Kitchen right on the premises. Baked goods don't get much fresher than that.

173 Sam Garner Road, Newport (Carteret County), 252-223-5283.
Open April to December.

Wilbur R. Bunch's Produce Stand

Everyone in Edenton knows "Bunch's," formally Wilbur R. Bunch's Produce Stand. The farm, fifteen minutes north of town in Rocky Hock, has

Not everyone is hard at work at Indigo Farms in Brunswick County.
Photo by Selina Kok.

In the spring, families flock to Garner Farms' produce stand in Carteret County
to pick strawberries. Photo by Selina Kok.

had a stand since 1988. The original stand abuts the farm, while a second,
larger one, opened in 2003 about five miles southeast of town. Run by
mother and son team Joyce and Keith Bunch (Wilbur died in 2008), the
farm itself has been going on "for years and years and years," said Joyce,
who lives in a 120-year-old farmhouse next to the Rocky Hock stand. The
farm is best known for its famed Rocky Hock melons, said Joyce, who also
puts up dozens of jars of pickles for sale. Being near the coast, Bunch's sells
some fresh seafood as well.

2833 Rocky Hock Road, Edenton (Chowan County), 252-221-4594.
Also at Highway 32 South at Soundside Road. Open May to November.

'R Garden

Third-generation farmer Julia "Kitty" Wethington likes to joke about how
she's "been left on the side of every road in Craven County with a trailer
full of vegetables and change money. Mom would say, 'I'll be back in an
hour or so to check on you.'" As Kitty notes, that was a different time.

Now she and her mother, Julie McKeon, sell produce off their sixty-five-acre farm in New Bern, established in 1947, as well as at several farmers' markets. Available are seasonal vegetables and bedding plants, including vegetables, herbs, and flowers. Kitty also cooks up and sells vinegar, jams, jellies, relish, preserves, and pickled vegetables. If you see her on the side of the road, wave hello.

605 South Glenburnie Road, New Bern (Craven County), 252-637-4172, www.rgardennb.com. Open late May to November.

West Produce

On a warm April day, we found Tommy West on his tractor, getting ready for the growing season at West Produce, the farm about ten miles north of Fayetteville that he owns with his wife, Jean. A third-generation farmer, Tommy turned his family's traditional tobacco farm into a produce farm in 1990. Sales are made at a farm stand that more resembles a market in its size and scope. "We have right much during the summer," Tommy said. That includes strawberries, okra, melons, potatoes, sweet corn, tomatoes, and squash. In September and October, the area becomes an agritourism destination, with barnyard animals, pumpkins, hay rides, and group tours for anyone from schoolchildren to church groups.

2026 Hayes Road, Spring Lake (Cumberland County), 910-497-7443.
Open April to November.

Rose Produce and Seafood Market

In 2007 Janet Rose and her husband, Paul, a commercial fisherman for three decades, opened Rose Produce and Seafood Market on Caratoke Highway on the way to the Outer Banks. Fish offerings from Paul typically include flounder, perch, and hard- and soft-shell crab, and from other locals he buys scallops and shrimp off the boat. While Rose, a second-grade teacher, had never been in retail, "we knew produce," she said. "Paul grew up on a farm in Currituck and I'm from Knott's Island," which has a history of farming. Janet goes out of her way to purchase local and regional produce, and Rose is typically the only stand around to carry peaches from North Carolina instead of South Carolina.

6378 Caratoke Highway, Grandy (Currituck County), 252-453-2911.
A smaller market is at the Virginia/North Carolina border on Caratoke Highway.
Open May to October.

Dail Family Produce

During the summer, Dail Family Produce in Snow Hill bustles with activity. Most produce is from Dail Family Farms. That includes beans, peas, blueberries, cabbage, cantaloupe, collards, cucumbers, onions, peppers, potatoes, squash, tomatoes, sweet corn, and watermelons. If you want, you can buy your peas and butterbeans shelled. You can also find fresh-cut flowers and honey at the popular stand.

4042 Highway 258 South, Snow Hill (Greene County), 252-560-8315.
Open June to August.

G. R. Autry and Son Farm and Peach Orchard

Former high school principle Raz McAutry first started growing peaches because he'd worked with local peach growers in the summers. "So I thought I knew a lot about peaches but found out I didn't," he said with a laugh. Two decades and thousands of peach trees later, McAutry has learned a few things, and now he looks forward to passing along his knowledge to his son George. Sales from their farm stand thirty minutes southwest of Fayetteville start in mid-June with produce. Throughout the summer the McAutrys grow okra, tomatoes, melons, sweet corn, sweet potatoes, and field peas. The eighteen acres of peaches are ready in July. Some customers come for the fruit and others for the ice cream. "We've been making peach ice cream for twenty years and sell a lot of it," Raz said.

1385 McGougan Road, Lumber Bridge (Hoke County),
910-875-3787. Open June to August.

Scott Farm Organics

Trent and Rebecca Scott are barely into their thirties, but they already are experienced farmers. Trent grew up farming tobacco, while Rebecca's family was in the logging business. Together since their teens, Trent and Rebecca used a Golden LEAF Foundation grant to help them start a certified-organic vegetable farm in 2007. On their 170 acres, which several family members work, they grow lettuce, cabbage, beets, radishes, sweet potatoes, and more, which they sell at their farm stand and the New Bern Farmer's Market. Trent's mother, who was helping at the roadside stand, noted, "We're going in a new direction, but there's things about farming that never change. It's hard."

Highway 17, one mile west of Craven/Jones County line, Rhems (Jones County),
252-229-2505, www.scottfarmorganics.com. Open spring to fall.

Carolina Country Fresh

The Roberson family was on top of things when the new Interstate 64 bypass was put in—they built their produce store between Raleigh and the Outer Banks next to the only gas station within ten miles in either direction. The Robersons have put a lot of work into their building inside and out, making this a popular stop for travelers. Outside, the rough-cut pine building with a red tin roof resembles a barn, and inside fruits and vegetables share space with old-timey decorations and knickknacks. The Robersons (four family members run the store, and two more run the farm) sell produce from their farm, as well as other local and out-of-state farms. In the spring they sell their own strawberries and blueberries and operate a three-acre U-pick strawberry patch one mile from the store. Our favorite item for sale was the plastic baggie filled with dried tobacco leaves. The store sells cotton, too, and, yes, people buy them both. "A lot of people say they've never seen them before," said Deann Roberson. "It's a novelty item."

707 North Main Street (Exit 502 off Highway 64), Robersonville (Martin County), 252-661-1002. www.carolinacountryfresh.com. Open April to December.

Johnson Farm

As times change, so do Johnson Farm's sales locations. Known mostly for its peaches, the Sandhills farm store is now in its fourth location, prompted by the addition of Interstate 73/74 and the Highway 220 bypass. Run by Garrett and Barbara Johnson and in business since 1934, the farm grows thousands of peach trees, as well as produce, on about fifty acres. Inside the market, whose bins are filled with just-picked peaches from June to September, is a gift shop with local art, pottery, and peach preserves, jams, and other products made with Johnson Farm's peaches. But the most popular feature is the snack stand, which sells homemade peach ice cream and peach dumplings, similar to turnovers.

1180 Highway 220 North, Rockingham (Montgomery County), 910-997-2920. Open May to December.

Auman Orchard

Peaches are to the Sandhills what apples are to Henderson County, and no other peach farm is as much an institution as Auman Orchard. It first flourished under the care of Clyde Auman, whose father started a small orchard in West End in the early 1900s. Clyde and his brothers expanded the orchards greatly, becoming some of the largest landowners in Moore County. During peach season, Clyde held court at the packing shed behind the family's home, chatting up customers and doing business. His son Watts, now in his seventies, carries on the tradition of both farming and conversation. From his home on the farm, Watts greets customers new and old as they drop by to pick up their bags or bushels of just-picked peaches. Even for folks who have been coming here for more than fifty years, the scenery, including the open-air shed, hasn't changed much. When asked if the orchard really is open on Sundays (most aren't), Watts replied, "We've tried to close, but the doorbell just keeps ringing."

3140 Highway 73, West End (Moore County), 910-673-4391.
Open June to September.

Chappell Peaches and Apples
Fourth-generation farmer Ken Chappell is rare in the Sandhills in that he grows several varieties of apples as well as the more typical local offering of peaches. From the family's basic farm stand along Highway 211 in Eagle Springs, Ken sells the fruits in season, as well as assorted vegetables and melons. In 2009 the farm added heirloom tomatoes to its mix.

672 Highway 211, Eagle Springs (Moore County), 910-673-1878
(works only during season), www.chappellpeaches.com. Open June to October.

Highlanders Farm

Seven generations of the Blue family have farmed in the Sandhills, with an eighth in line. The family arrived in 1804 from Scotland and has farmed a little of everything, including turpentine, cotton, tobacco, and now produce and strawberries. Siblings John Blue and Patti Burke and their spouses run the current operations, northeast of Pinehurst. In 2008 they opened a farm stand, gift store, ice cream shop, and U-pick strawberry patch. The shop is in a century-old building that had once housed the post office and railroad depot for the community, called Blues Siding. "The rail was for hauling turpentine and buggies," said Patti, who is responsible for gathering crafts "made by local ladies" to sell in her shop. Much of the produce for sale is from their farm, and the ice cream flavors include strawberry and peach from their own harvest.

5784 Highway 22, Carthage (Moore County), 910-947-5831.
Open April to September.

Kalawi Farm

From the day they open around Easter, Kalawi Farm's stand and the adjacent Ben's Homemade Ice Cream Shop stay busy. Art and Jan Williams started their peach orchard on family land in Eagle Springs in 1982 and added a stand in 1985. They grow about 5,000 peach trees on thirty acres, as well as row crops and produce, which also is for sale. Their thirty-five varieties of peaches are harvested through mid-September. Next to the stand under a grove of pine trees is an area with picnic tables for sitting and licking. The Williamses gave the farm its moniker using letters in the names of their children, Katie, Laura, and Will. When Ben came along several years later, he got the ice cream stand.

1515 Highway 211, Eagle Springs (Moore County), 910-673-5996.
Open April to November.

Bailey's Berry Farm

After Linda Bailey retired from thirty years of teaching at a Nash County elementary school, she found herself with too much time on her hands. "A friend suggested we grow strawberries," she said, so she and her husband Joe, a long-time tobacco farmer, opened a U-pick patch in 2001. "Then someone said, 'Why don't you sell cabbage, and maybe include spring onions, too.' Well, it just mushroomed from there." Linda and Joe, no longer in the tobacco business, grow a variety of produce on about fif-

Kalawi Farm in Moore County sells thirty-five varieties of peaches at its farm stand, and right out of the truck. Photo by Diane Daniel.

teen of their seventy-five acres. Their cute little market is housed in part of what was once a mule stable and sits behind their lovely century-old farmhouse. When we stopped by in June, the couple and several workers were working to fill a special order—sixty dozen cobs of sweet corn, shucked. And, yes, the shucking costs extra.

5645 Strickland Road, Bailey (Nash County), 252-235-4131. Open April to August.

Wrenn Farm

Old-timers might recall Middlesex Produce, started in Zebulon by three farmers in 1967 and later a fixture in grocery stores around the Southeast. "They were the first I know of in the state to grow greenhouse tomatoes," said Mitchell Wrenn, whose father was one of the original owners. Eventually only Wrenn Farm remained, and now it sells mostly retail out of its farm stand and at markets. Mitchell uses a long-discontinued hothouse tomato seed variety that he swears by. "When my father heard they were stopping it, he bought and froze enough seeds to last for decades." From his delightfully old-timey market, Mitchell sells vegetables from his farm only. He grows a bit of everything, but specialties are strawberries and butter beans. He also runs a U-pick strawberry operation. When we visited in June, we spied some whopper zucchinis, weighing in at more than four pounds. "People making bread, that's what they say they want, so we grow a few for them."

5078 Brantley Road, Zebulon (Nash County), 919-269-9781.
Open April to December.

Lewis Nursery and Farms

In the Wilmington area, the name Lewis is synonymous with berries. Farmer Cal Lewis runs a huge wholesale fruit and produce business, sending shipments near and far. In 1984 Lewis Nursery and Farms started a U-pick strawberry operation that in 1991 moved to its current address. Open from spring through early summer, it is wildly popular, so much so that the farm hires police officers to direct traffic on weekends. Prepicked and U-pick strawberries take up seven acres of the fifteen-acre farm, which grows blueberries and blackberries later in the season. A well-stocked plant nursery attracts the adults, while homemade ice cream keeps customers of all ages streaming in.

6517 Gordon Road, Wilmington (New Hanover County), 910-452-9659, 910-675-2394. Open April to July.

Nature's Way Farm and Seafood

When the Mollers met, Tina was working as a waitress in a seafood joint and Bill was a clammer. They married in 1983 and bought three acres near Topsail Island. Bill kept working the sea (crabbing and shrimping, too), Tina the tables, and they started growing vegetables. Tina discovered her passion after buying a few goats. Since 1990 she's worked full time on the farm, making a wide variety of cheeses from the milk of their dozen or so goats. Bill sold his shrimp boat, but he still goes crabbing and clamming. At the small shop next to their home, they sell fresh seafood and Tina's cheese and goat-milk body products. In 2009 a beekeeper brought eighteen hives to the farm, and Bill was pondering taking them over. "It's a new challenge," he said. And perhaps a new product for the store.

115 Crystal Court, Hampstead (Pender County), 910-270-3036. Open year-round.

Briley's Farm Market

If you love poking your head in those wooden goofy cutouts for silly looking photos, as we do, you'll find plenty of options at Briley's Farm Market in Greenville. Photo opportunities include a strawberry, school bus, and assorted barnyard animals. Oh yeah, and you can buy plenty of produce here as well from the family's sixty acres. We went during strawberry season, and the pick-your-own folks were out in full force. In the small, well-organized sales shack, there were berries, collards, onions, fresh herbs, and tomatoes for sale, as well as many potted herbs and flowers. To entertain the kids, there are a few farm animals, a playground, and, in the fall, hayrides, pumpkins, and corn mazes. Kids seem to like those cutouts, too, though we always thought they were for adults.

5290 Old Pactolus Road, Greenville (Pitt County), 252-754-5029,
www.brileysfarmmarket.com. Open May to October.

Renston Homestead

The packhouse, ice house, commissary, and other buildings from the Mc-Lawhorn family farm, dating from the early 1890s, are still standing today at the Renston Homestead in southern Pitt County. Originally the usual tobacco, corn, soybeans, and cotton were grown here, then later the family ran a dairy and beef cattle operation. In 2000 Steve McLawhorn and Mike Skinner planted produce and strawberries for the farm's most recent incarnation—a produce stand and U-pick strawberry operation (called Strawberries on 903). Along with corn, beans, melons, asparagus, and other vegetables, the farm specializes in potted and cut flowers. During the spring and fall, tours are offered to schoolchildren and adults, teaching them about both historic and current farm practices.

4064 Highway 903 South, Winterville (Pitt County), 252-321-3204,
www.renstonhomestead.com. Generally April to November.

The Berry Patch

When your family name is Berry and you're a farmer, it's only natural what you'd grow. Lee and Amy Berry took it even further, adding a U-pick operation, a produce stand, and an ice cream shop. Not just any old shop, but one shaped and painted like a strawberry, making it hard to miss along the highway, about seven miles north of Rockingham. The farm stand sells produce from the couple's farm as well as from regional farms and some out of state. Several flavors of homemade ice cream are made in wooden double churns, including local peach and, of course, strawberry when berries are in season. The shop also carries Granny Berry's jams, salsas, and pickled items made by Lee's mother.

1246 Highway 220 North, Rockingham (Richmond County), 910-895-6522.
Open March to November.

Triple L Farm

The charming market at Triple L Farm has a "staff" of about a dozen—all family members pitching in where needed. The seventy-year-old building was once a general store run by the grandfather of former tobacco farmers Jim and Joe Lambeth. "They sold everything from food and clothes to farm equipment," said Jim's wife, Marcia, who taught English in the Richmond County schools for thirty years. In 2002 she and others fixed up the Sandhills store, which had been used as storage for three decades. Outside are garden sculptures and flowers, and a U-pick strawberry patch is next door. Inside are bins of produce, mostly from Triple L Farm, along with local eggs and jams. On display are farms implements and old-time furnishings, games, and toys. "The store is just a lot of fun," Marcia said. "We all enjoy meeting up there."

2205 Derby Road, Ellerbe (Richmond County), 910-417-0438, www.derbystand.com. Second location at 1728 Highway 5, Aberdeen. Open April to August.

Geraldine's Peaches and Produce

Though the Sandhills are known for their peaches, a Lumberton farmer is proving that the lowlands of Robeson County are ripe to grow the fruit as well. Roy Herring and his wife, Geraldine, planted peach trees when they saw the tobacco buyout coming. In 2005 they were ready to sell their first harvest, and now they have about 2,500 trees, some open to U-pickers. At their roadside stand, Geraldine's Peaches and Produce, they sell the bounty from their diverse produce garden. "Everything for sale here is from our farm," Geraldine said. Items include blackberries, raspberries, melons, potatoes, tomatoes, okra, and greens. While the farm is known for its produce, it also had its fifteen minutes of fame as the location for several scenes in the 2008 Hollywood film *The Secret Life of Bees*. But because the filming had to take place in the winter the leaves you see on screen are made of silk and the fruit is plastic.

10728 Highway 41 North, Lumberton (Robeson County), 910-739-8686.
Open May to December.

FARMERS' MARKETS

Sandhills Farmers Market

Ammie Jenkins remembers the impromptu "farmers' markets" of her childhood. "Everyone would bring what they'd grown and my grandfather would tell stories." That's the sort of scene she encourages at the Sandhills Farmers Market in Spring Lake, just north of Fayetteville, billed as the state's only African American market. "People come and sit down and they talk and get to know their growers. It's not a big market, but we try to give it a down-home feel," said Ammie, executive director of the Sandhills Family Heritage Association, a nonprofit group that helps preserve the traditions of African American families in the region. The market's cinder-block building was constructed by the community as a gathering place during segregation and later was used for meetings during the Civil Rights era.

230 Chapel Hill Road, Spring Lake (Cumberland County), 910-436-3406.
Held Saturday mornings June to October.

Moore County Farmers Market

The Moore County Farmers Market has been operating at various locations since the 1970s and now has multiple sites. In early June we visited the Saturday market in historic downtown Southern Pines. A dozen vendors were set up on Downtown Park, the former site of the police station, while musicians performed and people milled about shopping and showing off their dogs (leashed, of course). The offerings included just-picked strawberries and peaches, as well as other produce, eggs, honey, flowers, herbs, and baked goods. All Moore markets are "producer only," with goods coming from within fifty miles.

100 Southeast Broad Street (at New York Avenue), Southern Pines (Moore County), 910-947-3188, www.localharvest.org. Held Saturday mornings April to September.

Nash County Farmers Market

Like many cities, Rocky Mount had a curbside downtown farmers' market for decades. Twenty years after it closed in 1970, a group of farmers started to meet up and sell produce on Saturdays on the grounds of a local church. After some lobbying and the awarding of county and state grants, a permanent enclosed market was built in 2005. Now, the Nash County Farmers Market attracts hundreds of customers a week and has added a second building. We visited on a slower weekday in June and were surprised to find not only several produce vendors but also a musician performing country ballads. While vendors here are allowed to sell products from other farms as well as their own, all produce must come from North Carolina.

1006 Peachtree Street, Rocky Mount (Nash County), 252-407-7920.
Held Saturdays mornings April to November and one weekday (varies) June to August.

Farmers' Market at Poplar Grove

Understandably, weekday farmers' markets are typically smaller than their Saturday counterparts or standalones. The Farmers' Market at Poplar Grove Historic Plantation is a notable exception. On Wednesdays during season, this former peanut plantation, now a nonprofit historic site, returns to its farming roots. Some thirty-five vendors set up on the expansive lawn under big shade trees in the shadow of the 1849 manor house. Products for sale include produce, seafood, flowers, goat cheese, and meat, as well as some arts and crafts. While you're there you can stroll the grounds for free or tour the house and exhibits for a fee. The market starts up just after the plantation's popular Herb and Garden Fair the last weekend in March.

10200 Highway 17 North, Wilmington (New Hanover County), 910-686-9518, www.poplargrove.com. Held Wednesdays mornings April to December.

Saving Our Animals, Breed by Breed

Modern food production is whittling down the number of both vegetable and livestock species, favoring only those offering maximum output in a controlled environment. Because of this, many traditional livestock breeds have lost popularity and are threatened with extinction. The American Livestock Breeds Conservancy (www.albc-usa.org), founded in 1977 and based in Pittsboro, works to conserve rare breeds and genetic diversity. These traditional breeds are not only essential parts of the history of American agriculture, the flavors from their meat, from Tamworth pigs to Delaware chickens and Bourbon Red turkeys, are richer and tastier.

Riverfront Farmers' Market, Wilmington

The Riverfront Farmers' Market in Wilmington has one of the most scenic market locations in the state. Vendors line up along two blocks of the mile-long Riverwalk with a full view of the Cape Fear River. The walkway, which parallels downtown Wilmington, is already the city's top tourist attraction, so with the addition of the market, which includes live entertainment, Saturdays are hopping. The market, run by the city of Wilmington, opened in 2004 and has continued to expand. You'll find about twenty-five farmers and ten craft vendors, selling everything from fruits, plants, eggs, cheese, and baked goods to, of course, fresh seafood.

North Water Street at Market Street, Wilmington (New Hanover County), 910-538-6223, www.wilmingtonfarmers.com. Held Saturday mornings April to December.

Onslow Farmers Market

When a pie baker at the Onslow Farmers Market in Jacksonville offered to make any variety we wanted for the next week, we pondered how to lure her to our home market in Durham. The twice-weekly Onslow Farmers Market, operating since 1996, houses twenty to thirty vendors in its covered building behind the Onslow County Cooperative Extension Service. Most are conventional farmers, with a standout exception being Whispering Dove Goat Ranch and Apiary. Vendors have to grow only half of what they sell, meaning you don't always meet the farmer. There's often entertainment here, with the most popular event being the annual talent contest, which usually draws a couple thousand relatives, um, spectators.

4024 Richlands Highway, Jacksonville (Onslow County), 910-455-5873.
Held Tuesday and Saturday mornings April to October.

Downtown Waterfront Market

When Elizabeth City leaders decided to start a downtown farmers' market, they wisely incorporated the area's best asset — the Pasquotank River. Started in 2008, the Downtown Waterfront Market quickly became popular. Vendor tents are set up on the lawn of Mariners' Wharf Park, which looks directly onto an open expanse of water. A small stage sets the scene for chef events and musical entertainers. Products for sale must be grown or made by the vendors, with a few exceptions. Typically about thirty produce and craft tents dot the park during high season. When you're finished shopping here, historic Main Street, just across the street, beckons.

100 block of South Water Street, Elizabeth City (Pasquotank County), 252-338-0169, www.downtownwaterfrontmarket.com. Held Saturday mornings May to October.

CHOOSE-AND-CUT CHRISTMAS TREES

Moore Christmas Tree Farm

Ricky Jones, who has a forestry degree, has worked with wood most of his life, most recently buying and selling timber for the giant Weyerhaeuser Company. In 1985 he helped his father-in-law start a Christmas tree farm, which Ricky has since taken over. "It's really more a hobby for me," said Ricky, whose extended family pitches in at the ten-acre Moore Christmas Tree Farm during the holidays. "My nieces and nephews probably make more money at it than I do," he joked. Ricky grows Leyland cypress, Virginia pine, white pine, red cedar, and Carolina sapphire. And, yes, precut Fraser firs are on hand for those in need. "We make wreaths and garland, too," he said, "but the big thing you're selling is the adventure."

405 Parker Road, New Bern (Craven County), 252-638-4160.
Open late November to December.

Beautancus Christmas Tree Farm

Not to worry. If you're challenged by the pronunciation of Beautancus Christmas Tree Farm, the owners have printed the phonetic version right on their sign: Bo-tank-us. "That was my mother's and my idea, because no one knows how to pronounce it," said Suzanne Southerland, who with her husband, Brownie, owns the farm on his family's land near Mount Olive. The choose-and-cut farm, a retirement project of Brownie's, a former state park ranger, started in 1984. Brownie grows five acres of red cedar, Leyland cypress, and white pine, and ships in Fraser firs. The farm is located off a picturesque rural road also called Beautancus in the community of the same name. Good thing we know how to pronounce it.

1569 Beautancus Road, Mount Olive (Duplin County), 919-658-4512, 919-778-7060. Open late November to December.

Doby Christmas Tree Farm

Donny Doby and his wife, Carol, are known as first-rate tree growers who farm full time, growing produce, row crops, hay, and wheat straw. Their four acres of trees include white pine, Virginia pine, red cedar, and Leyland cypress. Only in 2008, after nineteen years of selling his trees, did Donny start bringing in Fraser firs from the western part of the state. "Some people, it's all they want," Donny said. If you visit, be forewarned that Donny's road changes names halfway through, from Doby to Edmunds. That's because Donny's grandfather, a Doby, and a neighbor, an Edmunds, couldn't agree on what to call it, so they each took an end.

150 Doby Road, Cameron (Moore County), 910-245-3265. Open late November to December.

VINEYARDS AND WINERIES

Lu Mil Vineyard

Within a year of opening Lu Mil Vineyard near Elizabethtown in 2004, owner Ron Taylor was expanding it. Named in honor of his parents, Lucille and Miller, the winery began in an old barn. By 2005 Ron had upgraded to a new, larger building with a gift shop and events center. Most of the thirty-eight acres here are filled with tidy rows of muscadine vines (with twenty more acres grown nearby), but Ron didn't stop there. In 2008 he added three rental cabins (with plans for more) and is opening an RV park. A pond on the grounds welcomes anglers and paddle boaters. A deli, opened in 2009, draws a lunch crowd, many of which sits on the deck overlooking the vineyards. The winery does a brisk wedding business as well. And in December Lu Mil is aglow with some 350,000 Christmas lights, drawing folks from near and far for its annual Festival of Lights.

438 Suggs-Taylor Road, Dublin (Bladen County), 910-866-5819, www.lumilvineyard.com. Lodging $$

Grapefull Sisters Vineyard

When sisters Amy Suggs and Sheila Suggs-Little were thinking about what to do with the fifty acres of family land they inherited, they cursed those annoying muscadine grapevines, always taking over everything. In 2006 they made their enemy their friend, opening Grapefull Sisters Vineyard, where they cultivate five acres of grapes in soil that used to feed cotton. The property, which has been in the family for eight generations, is northwest of Myrtle Beach. Their wood-frame building holds a gift shop and tasting room downstairs and Inn d'Vine, a bed and breakfast, upstairs. Also on the property is Grapefull's small RV park, CarrollWoods, which opened in 2009. The sisters also are planning to open a small bistro. When we visited, beach-going tourists from Ohio were receiving their first wine-tasting from Sheila and clearly enjoying the lesson.

95 Dot's Drive, Tabor City (Columbus County), 910-653-2944, www.grapefullsistersvineyard.com. Lodging $$

Duplin Winery

Duplin Winery is the perfect stop for anyone on the way to beaches via Interstate 40 and those intimidated by wineries, two groups that, combined, probably include half the state's population. Indeed, North Carolina's highest-volume winery gets close to 100,000 visitors yearly and sells more than 250,000 cases of its award-winning muscadine wines. Founded by brothers David and Dan Fussell and now run by David Fussell Jr., Duplin sold its first bottle in 1975 and has grown wildly since. Duplin has 100 acres of vines on site and oversees production of about 1,500 acres in North Carolina and three neighboring states. The winery operates a large gift shop, bistro, and even a dinner theater, and it gives vineyard tours. But the liveliest action happens along its forty-four-foot tasting counter, where customers partake in free samples from the winery's more than twenty-five labels. Two of Duplin's best-known are the multiple-award-winning

Magnolia and the Mothervine, a scuppernong produced with cuttings from the 400-year-old native Mother Vine on Roanoke Island, believed to be the world's oldest cultivated grapevine.

505 North Sycamore Street, Rose Hill (Duplin County), 800-774-9634, www.duplinwinery.com.

Bannerman Vineyard/White Oak Farm

The moment we arrived at Bannerman Vineyard in Burgaw on a steamy day in July, our pulse slowed. Vineyards surrounded us as far as the eye could see, the country road out front was blessedly quiet, and the old to-bacco barn next to the tasting ring was artfully ringed with muscadine vines. The small winery, which opened in 2003, is run by Scot and Colleen Bannerman and their son, Chris. The eighty-acre farm has been in the family since 1971, when Scot's father bought it, calling it White Oak Farm. He ran it as a U-pick muscadine vineyard, and the sixteen acres in produc-tion are still open for public picking. "I remember as a kid sitting under the awning there in the shade and weighing grapes with a mammoth scale," Scot said, pointing to the tobacco barn. The tasting room is attractive and comfy, but the most inviting spots are outside. Scot agrees, adding, "I like for people to come sit in the rocking chairs." His wish is our command.

2624 Stag Park Road, Burgaw (Pender County), 910-259-5474, www.bannermanvineyard.com. U-pick September to October.

STORES

Bertie County Peanuts

After a decade of working as an athletic director at a Cary prep school, Jon Powell returned home to Bertie County in 2007 to help run the family business. Since 1919 Bertie County Peanuts has been selling raw peanuts from local farmers for processing, and in 1992 it started a retail business, which it expanded nationally in 2009. Customers can stop by the family's farm supply company, Powell & Stokes, in sleepy Windsor, for samples and sales of products from peanut butter to our favorite, blister fried nuts. The peanuts come from local and Virginia farms, Powell said, and though the processing is done in Virginia, the products are made at the company's cookhouse a mile away. If you ask ahead, in September or October, Powell will arrange a tour of the peanut-processing plant.

217 Highway 13 North, Windsor (Bertie County), 252-794-2909,
800-457-0005, www.pnuts.net.

Lockwood Folly Marketplace

Only three years after graduating from college, Lindsay Hewett, along with her husband, Travis, opened Lockwood Folly Marketplace in Supply. The town, ten miles north of Holden Beach, was named for the many supplies that once came up the Lockwood Folly River by boat. In 1943 Lindsay's great-grandparents opened a general store there, moving to a larger building in 1963. Lindsay and Travis opened their own shop and deli in the same location and are working to create a similar community feel. "We're really focusing on local farms and artisans," said Lindsay, who previously worked as a county extension agent. She stocks gift and food items from Brunswick County and throughout the state, with an emphasis on those naturally made or grown, including produce, free-range meat, artisanal cheese, and hormone-free milk.

48 Stone Chimney Road, Supply (Brunswick County), 910-754-5445,
www.lockwoodfollymarketplace.com.

Mackey's Ferry Peanuts

Mackey's Ferry Peanuts, open since 1983 and a traditional stop for beach-goers, has seen a lot of changes in the last decade. The Smith family bought the shop in 2003, then two years later moved it twenty-two miles west, between Plymouth and Jamesville, when the Highway 64 bypass rerouted traffic. But as for what its customers care most about—the peanuts and peanut products—"we pretty much kept the business the way you've always seen it," Sharon Smith said. The name Mackey's Ferry comes from the ferry once used by the Norfolk Railroad to carry train cars across the Albemarle Sound. Using peanuts grown in North Carolina and Virginia, the Smiths make shelled crunchy peanuts, peanut brittle, peanut butter, and more. They also keep a seasonal demonstration peanut patch going so folks can see the legumes from start to finish. "We have lots of people wanting to dig them up to see what they look like in the ground," said Sharon, who gladly obliges.

30871 Highway 64 East, Jamesville (Martin County),
888-637-6887, www.mfpnuts.com.

Tidal Creek Cooperative Food Market

Like most co-ops, Tidal Creek Cooperative Food Market has a nice community atmosphere, although its location a few miles from downtown doesn't necessarily appeal to visitors. The store, which started in 1982 in a smaller location, is bright and airy, with a good supply of local produce during the growing season and cheese and herbs year-round. Monthly events, occasional cooking classes, and farm tours add to the communal spirit.

5329 Oleander Drive, Wilmington (New Hanover County),
910-799-2667, www.tidalcreek.coop.

B&B Pecan Processors of North Carolina/Elizabeth's Pecans

Located three miles west of Interstate 40, Elizabeth's Pecans outlet store is a magnet for holiday travelers in search of Carolina gifts. The nondescript shop sits in front of B&B Pecan Processors of North Carolina, where the pecans are shelled and the candy is made. The nuts are harvested around the corner from a forty-five-acre grove planted in 1981 by retired NASA engineer Bobby F. Bundy. His son, Alan, took over after his father died. Alan later added a candy-making component, named for his daughter, Elizabeth. Today, he harvests about 38,000 pounds of pecans on producing years (pecan trees yield every other year). The top-selling candy (most sales are online) is Elizabeth's Exceptional Pecan Brittle, which is easier on the teeth than traditional peanut brittle and has a smooth, buttery flavor. Alan also has planted a peach orchard behind the pecan trees and uses the fruit in specialty items, including preserves and salsa.

107 Thomson Avenue, Turkey (Sampson County), 910-533-2229, 866-328-7322,
www.elizabethspecans.com. Tours by appointment.

DINING

Weeping Radish Farm Brewery

If your idea of a quality suds and dog is a Bud Light and an Oscar Mayer wiener, Weeping Radish Farm Brewery has a cure for that. The state's oldest microbrewery has made unfiltered, natural beer since 1986. In 2006 German owner Uli Bennewitz moved the business from Manteo to a twenty-four-acre location along well-trafficked Caratoke Highway between Virginia and the Outer Banks. Uli left the Bavarian restaurant theme behind for a farm-to-fork approach, which includes a brewery, a butcher, meat

Uli Bennewitz of Weeping Radish Farm Brewery in Currituck County explains
the sausage-making process. Photo by Diane Daniel.

and produce sales, and a restaurant. Weeping Radish's sausages and other
meat cuts, which you can sample at the brewery restaurant or purchase
uncooked, are made by an on-site German butcher using pork from local
sustainable farms and cooked in the local smokehouse. Uli's fourteen-acre
farm supplies the restaurant's produce, with enough left over for retail
sales. "My goal at the restaurant is to have no menu and only use what we
grow at the time," he said. To that we say, "Prost!"

6810 Caratoke Highway (Highway 158), Jarvisburg (Currituck County),
252-491-5205, www.weepingradish.com. $–$$. Weekly tours.

The Sanderling Resort and Spa

When we caught up with chef Joshua Hollinger in October, he had just
finished making fifty quarts of pesto from sixty pounds of basil before the
growing season ended. Since Joshua took the job of executive chef at the
tony Sanderling Resort in Duck in 2008, the two restaurants there have

experienced a sea change. "My basic mandate is buying as much local and sustainable as much as possible." Although Joshua came from another waterfront resort, the Harbor View Hotel on Martha's Vineyard in Massachusetts, there is a big difference between the two locations. The Vineyard, an island in the Atlantic Ocean, is covered with farms. On the Outer Banks, "I haven't found a farm to deliver here." Nonetheless, Joshua goes way out of his way to work with small farmers on the coast and the Piedmont. When we left him, he was looking to contract with a goat farmer for milk to make his own cheese.

1461 Duck Road, Duck (Dare County), 800-701-4111, 252-261-4111, www.thesanderling.com. $$$$

Chef & the Farmer

When Chef & the Farmer opened in Kinston in 2006, it was big news here. Couple Vivian Howard and Ben Knight brought progressive dining to a formerly culinarily deprived city. The downtown restaurant also brought work, not only to its employees but to the growing list of coastal farmers whose products it uses. Vivian, the chef, and Ben, the manager and artist in residence, moved south from New York City because Vivian, a native of nearby Deep Run, received an offer from her family she couldn't refuse: come home and start a restaurant. The results are noteworthy. Menu standouts have included pumpkin risotto, butterbean hummus, and chicken breast stuffed with fig, caramelized onion, and goat cheese. Welcome back, Vivian.

120 West Gordon Street, Kinston (Lenoir County), 252-208-2433, www.chefandthefarmer.com. $$-$$$

Ashten's Restaurant

One of the long-standing favorite appetizers at Ashten's Restaurant in historic downtown Southern Pines is "Artichoke Heart Pops," artichokes stuffed with herbed chèvre from Goat Lady Dairy in the Triad, local tomatoes, and fresh herbs. Chef and owner Ashley Van Camp favors heritage pork from Cane Creek Farm, oysters from the North Carolina coast, and poultry and produce from around the Sandhills. For those scared off by the somewhat pricey menu, Ashten's also cooks up moderately priced pub fare served in its cozy bar, and just as locally sourced.

140 East New Hampshire Avenue, Southern Pines (Moore County), 910-246-3510, www.ashtens.com. $$-$$$

Where to Find a Real Catch

Many diners are unaware that the "fresh seafood" they're being served, even in our coastal towns, often comes from another country. Nationwide, about 84 percent of the seafood we eat is imported. To promote the local bounty in Carteret County, which includes the towns of Beaufort, Morehead City, and Atlantic Beach, fishermen, researchers, and supporters in 2005 started the marketing program Carteret Catch (www.carteretcatch.org). The brand label, seen in the windows of participating restaurants and seafood retailers, signifies that the seafood labeled as local comes directly from Carteret County fishermen.

In 2009, Carteret Catch fishermen became the starting point of supply for Walking Fish (www.walking-fish.org), a "community-supported fishery" started by graduate students at Duke University's Nicholas School of the Environment. Based on the CSA model, the Triangle's first CSF supplies area residents with locally caught seafood.

Catch

Catch started out in 2006 as a downtown neighborhood seafood joint, a twenty-two-seat restaurant with limited hours and usually a wait for a table. When chef Keith Rhodes opened a second, larger location in 2010, his fans multiplied. His inventive Asian-tinged menu declares: "Seafood is a natural product; availability is limited due to Mother Nature." Keith works with several area farms for produce, buys some local meat, and all local seafood. "We try to be very seasonal and local," Keith said. "For produce, herbs, seafood, and vegetables, we try to source no further than 100 to 120 miles." Keith has been pleased that, in an area where many restaurants rely on imported seafood, "the community has really supported us and our philosophy."

215 Princess Street, 910-762-2841 and 6623 Market Street, 910-799-3847, Wilmington (New Hanover County), www.catchwilmingtonnc.com. $$

LODGING

Benjamin W. Best Inn and Carriage House

When Mary Betty Kearney learned that the historic Benjamin W. Best House was going to be destroyed, she had to do something to save it. So she and her husband, Ossie, bought the run-down Greek Revival house, built in 1850, and moved it to their sprawling cattle farm southwest of Snow Hill. Their grass-fed beef, marketed under the name Nooherooka Natural, is sold in stores and from the farm. After renovating the house, now on the National Register of Historic Places, the couple opened the Benjamin W. Best Inn and Carriage House. Guests are welcome to take a golf-cart tour of the ninety acres and learn how the Kearneys raise their cattle. Weddings have become popular here, with the ceremony taking place on the front lawn, whose drive is lined with magnolia trees. A formal balcony graces the front of the house. "When we have weddings here, they love this balcony," Mary Betty said. "It's like Buckingham Palace."

2029 Mewborn Church Road, Snow Hill (Greene County), 866-633-0229, 252-747-5054, www.bwbestinn.com. $$

Big Mill Bed and Breakfast

The way Chloe Tuttle describes her childhood at the site of Big Mill Bed and Breakfast, you can just imagine her running around playing on the former tobacco farm. When she grew up, she headed for the city lights, traveling the country and beyond for decades. When she inherited her fifty-acre home place in 1984, she couldn't bear to see it sold. "I came back because I'm sentimental," said Chloe, whose brother still farms on some of the acreage there. Now she brings the world to Williamston at her beautifully decorated B&B. In each of the four self-sufficient units, two in the main house and two in the old tobacco packhouse, nostalgic touches share space with contemporary design and amenities. Chloe also delivers each guest a wonderful breakfast, using local eggs, fruit, berries, and pecans from her trees, often mixing them into her yummy homemade granola.

1607 Big Mill Road, Williamston (Martin County), 252-792-8787, www.bigmill.com. $$–$$$

Springfield Bed and Breakfast

At the Springfield Bed and Breakfast, Mary White jokes that she runs a convalescent home for aging chickens. "I can't do the chop-chop thing," she said. Over the years, she and her husband, Joe, also have taken in a blind mule, an aging horse, and two goats on the sixty-acre Hertford farm. The land has been in Joe's family since 1891. They raise a small number of Angus cattle, and their son farms twenty-three acres of rotating row crops next to the house, which sits just off Highway 17. The centerpiece is the White's handsome 1895 farmhouse, which the couple painstakingly restored in 2002. Mary, who enjoys showing guests the animals and vintage farm equipment stored in the barn, also serves up a serious country breakfast.

962 South Edenton Road, Hertford (Perquimans County), 252-426-8471, www.springfieldbb.com. $$

Jackson Farm

"What I'm passionate about is preserving things," said Jan Mann, who with husband Tom Jackson moved from Raleigh in 1980 to the 1800s family farm Tom grew up on. Not only has Tom, a former English teacher and journalist, rebuilt or renovated many structures on the property, he took a dilapidated cabin by a pond and transformed it into a beautiful rental cottage. Jan, a longtime potter, has a studio on the grounds, and husband and wife work their small farm, where they grow specialty greens, asparagus, and native plants. Tom is fixing up another guest house and plans to build a third structure near the wetlands. The couple has placed much of their 300 acres in a conservation easement. The cottage on the pond sits along the North Carolina Birding Trail, which includes a mile-long walking path through the farm and over a dam topped by a small screened porch, a perfect setting for the many wedding vows exchanged here.

13902 Dunn Road, Godwin (Sampson County), 910-567-2978, www.jacksonfarm.com. $$

SPECIAL EVENTS AND ACTIVITIES

North Carolina Strawberry Festival

Until early May, not a whole lot happens in Chadbourn, a town of 2,000 people fifty-five miles west of Wilmington. Then come the visitors, more than 10,000 of them, for the three-day annual North Carolina Strawberry Festival, the state's oldest and largest agricultural festival. (The state ranks fourth in the nation in strawberry production.) The fair started in 1932,

You Say You Want to Work on a Farm?

Many farms in North Carolina and beyond offer internships and apprenticeships. Expect to do grunt work for minimal pay, with the opportunity to learn plenty. Housing may range from a room in the farmhouse to a camper, and the time commitment typically runs between one and four months.

Two great resources for positions are posted online through the National Sustainable Agriculture Information Service (ATTRA), open to all, and World Wide Opportunities on Organic Farms (WWOOF), available to members, who pay a nominal yearly fee.

The WWOOF movement (www.wwoof.org), which began in 1971 in the United Kingdom and is now global, has been criticized as being often used as a cheap lodging option for young travelers, but members include hard-working farmhands as well. ATTRA (www.attra.org) tends to attract more serious applicants, according to farmers.

If you want to stay home and keep your hands clean, you can always become a virtual farmer over at FarmVille (www.farmville.com).

when Chadbourn shipped out hundreds of trainloads of berries. Events include a 150-unit parade, music, crafts, and food vendors. Berries are celebrated in the strawberry-spitting and strawberry-hat-decorating contests, as well as in a berry quality contest among farmers. And don't worry, organizers promise enough strawberry shortcake to go around, and then some.

Downtown Chadbourn (Columbus County), www.ncstrawberryfestival.com. Held in May.

North Carolina Muscadine Harvest Festival

The largest wine festival in the eastern part of the state is the North Carolina Muscadine Harvest Festival in downtown Kenansville. This official homage to the Southeastern native grape started in 2005 and attracts more vendors and crowds yearly. Festivities include muscadine cooking and wine-making contests, exhibits of old and new grape-harvesting equipment, educational seminars, music, food, crafts, and, of course, an ample opportunity to sample the goods from some twenty wineries.

Downtown Kenansville (Duplin County), 910-290-1530,
www.muscadineharvestfestival.com. Held in September.

Halifax County Harvest Days Festival

This annual agricultural festival southeast of Roanoke Rapids is made all the more interesting for where it's held, at the county's 4-H Rural Life Center. Set on an impressive 345 acres not far off Interstate 95, the center holds several buildings from the early 1900s. The main building started as the county home for indigent residents, then became a nursing home, and now is used for 4-H camps. On the grounds are a dairy barn, a corn crib, a furnished historic farmhouse, the county agricultural museum, and the restored two-room Allen Grove Rosenwald School, the first school in the area for black students, opened in 1922. Harvest Days activities include craft and farm demonstrations and an antique tractor pull. The facility also has horse stalls, trails, and camping for riders.

13763 Highway 903, Halifax (Halifax County), 252-583-1821, www.visithalifax.com.
Rural Life tours by appointment. Festival held in October.

Richlands Farmer's Day

Though it's held just northwest of the city of Jacksonville, Richlands Farmer's Day is 100 percent country. Held at Richlands High School, the event is best known for its contests: collard cooking, frying-pan toss, watermelon eating, egg toss, sack races, and tobacco spitting (you must be eighteen to enter, of course). Instead of the usual beauty contest, there's a Lil' Mr. and Miss Farmer's Day contest for children four to six, who dress up as farmers or farmers' wives (what about farmers' husbands?). During the "Parade of Power," antique tractors on display all day are revved up and chugged by the crowd. All proceeds go to the Friends of Farmers scholarship fund, which assists in educating future farmers.

8100 Richlands Highway (Highway 258), Richlands (Onslow County),
910-324-7492, 800-932-2144. Held in September.

North Carolina Blueberry Festival

Though the inaugural North Carolina Blueberry Festival wasn't staged until 2004, the first cultivated blueberry production in Pender County started in the 1930s. With its ideally sandy soil, the county is second only to Bladen in blueberry production in the state, making it Pender's most important crop. With the advent of the annual festival, which draws 20,000 to charming downtown Burgaw, the berry has become part of the county's public identity as well. The party starts on Friday with a golf tournament, a barbeque cook-off, and a blueberry recipe contest, and it continues through the weekend with music, crafts, parades, and of course all the berry products you could hope for, from lip balm to muffins to smoothies. Just think of all the antioxidants in the air.

Downtown Burgaw (Pender County), 910-259-9817, www.ncblueberryfestival.com. Held in June.

Ayden Collard Festival

Collards are one of those leafy greens that folks most love or hate. Some won't even try them, while others can't get enough. (We fall somewhere in the middle.) Since 1974, the small town of Ayden, just south of Greenville, has celebrated the collard with its Annual Collard Festival. In 1984 organizers even sold a booklet called *The Collard Poems*. The collard-eating and collard-cooking contests remain popular, featuring such delicacies as collard pizzas, pies, and tacos.

Downtown Ayden (Pitt County), 252-746-2266, www.aydencollardfestival.com. Held in September.

RECIPES

Summer Squash Cakes

These squash cakes are a summertime favorite at the home of 'R Garden farmer Kitty Wethington in New Bern. While she doesn't sell these, Kitty and her mother, Julie, do make jams, salsas, and more that they sell at several farmers' markets.

MAKES 12 3-INCH CAKES

4	medium summer squash, yellow, green, or a combination of both (about 2 pounds)
1	small yellow onion
1	small russet potato, peeled and coarsely grated
	Kosher salt

1	clove garlic, minced
1	tablespoon fresh thyme or oregano, finely chopped
1	egg, lightly beaten
⅓–½	cup all-purpose flour
¾	cup grated Parmesan cheese (about 3 ounces)
	Freshly ground black pepper
6	tablespoons canola oil

Grate the squash, onion, and potato separately with a hand grater. Place the squash in a colander or strainer and sprinkle with salt. Allow to sit for 5 minutes (do not let it sit longer or it will get too soft), rinse under cold water, and carefully press out the water. Place the onion in a strainer and press out the water.

In a large bowl combine the squash, onion, potato, garlic, thyme, egg, flour, cheese, and pepper. Add enough flour to make a batterlike substance that is fairly thick. Heat 2 tablespoons of the canola oil in a large nonstick griddle or skillet over medium heat. Place 3 tablespoons of the batter on the hot griddle and press lightly to make a 3-inch round. Cook on one side until golden brown, about 4 to 5 minutes. Turn and cook on the other side. Remove and keep warm in a 250-degree oven. Repeat with the remaining batter, adding more oil as necessary.

Big Mill Granola

When we visited Chloe Tuttle at Big Mill Bed and Breakfast in Williamston, she had just finished making a big batch of homemade granola, which sometimes even features pecans from her own trees. After devouring a bowl at breakfast, we had to have the recipe for this crunchy treat that fueled our day.

MAKES 9 CUPS

5	cups old-fashioned whole oats (not quick-cooking) (about 1 ¼ pound)
2	cups whole pecans (about 8 ounces), slightly chopped
½	cup flax seeds (about 3 ounces)
1 ½	teaspoon ground cinnamon
1	teaspoon freshly grated nutmeg
	Pinch of salt
⅔	cup honey
⅓	cup canola oil, plus more for greasing the baking sheet

<blockquote>

½   cup freshly squeezed orange juice (from one large orange)

2   teaspoons vanilla extract

1 ½   cups dried cranberries, dried blueberries, or raisins
</blockquote>

Preheat oven to 350 degrees. Liberally grease a large-rimmed baking sheet, such as a jelly roll pan.

In a large mixing bowl, combine the oats, pecans, flax seeds, cinnamon, nutmeg, and salt. In a small saucepan combine the honey, oil, and orange juice. Cook over low heat until the honey has melted, about 1 to 2 minutes. Remove from the heat and stir in the vanilla. Pour the liquid over the dry ingredients and stir well until the oat mixture is moistened.

Spread the granola on the greased baking sheet and bake for 15 minutes. Remove from the oven and stir, then continue to bake for 20 minutes, stirring halfway through. As the granola begins to brown, stir every 5 minutes. It may take 45 minutes to 1 hour to cook. The granola is done when it is golden brown.

Remove from the oven and immediately transfer to a bowl and allow to cool. Stir in the cranberries, blueberries, or raisins. Store in an airtight container.

Sweet Potato Pie Ice Cream

Not only do Jackie and Louie Hough of Raft Swamp Farms raise sweet potato plants from the sweet potatoes they grow, cure, and save through the winter, they even make their own sweet-potato ice cream. You'll love it.

MAKES HALF A GALLON

<blockquote>

1 ½   cups milk (at least 2 percent; do not use skim milk)

4   egg yolks

1   cup honey

½   teaspoon cinnamon

½   teaspoon allspice

¼   teaspoon ground ginger

¼   teaspoon salt

1 ½   cups sweet potatoes (about ¾ pound to 1 pound), baked, peeled, and mashed

½   teaspoon vanilla

3   tablespoons dark or amber rum

2   cups heavy whipping cream
</blockquote>

In a medium heavy-bottom saucepan, whisk together the milk, egg yolks, and honey until combined. Cook over medium-low heat, stirring constantly with a wooden spoon until thick enough to coat the back of the spoon, about 10 minutes. You will know it's ready when you can see where you've run your finger down the back of the custard-coated spoon. Do not allow the custard to boil. Pour into a large bowl.

Whisk in the remaining ingredients until thoroughly blended and allow to cool. Freeze the custard in an ice-cream maker according to the manufacturer's instructions. Store in an airtight container.

Pan-Seared Rockfish with Sweet Onions and Sweet Potato Puree

Since landing at the posh Sanderling Resort and Spa on the Outer Banks, executive chef Joshua Hollinger has made it his mission to fill the menus of his two restaurants with dishes based on local and sustainable ingredients. Fittingly, this magnificent meal shines the spotlight on bounty from North Carolina's sea and soil.

SERVES 4

SWEET POTATO PUREE

5	sweet potatoes (about 3 pounds)
4	tablespoons unsalted butter, melted
½	cup heavy cream
1	tablespoon fresh chives, thinly sliced
1	tablespoon fresh parsley, finely chopped
1	tablespoon fresh thyme, finely chopped
	Salt and freshly ground black pepper

Preheat oven to 400 degrees.

Prick the sweet potatoes all over with a fork, wrap in aluminum foil, and bake for 1 hour until tender. Remove them from the oven and let cool slightly. Scoop the flesh from the skin and puree in the bowl of a food processor until smooth. Transfer the sweet potatoes to a medium bowl. Add the butter, cream, and herbs, and stir to combine. Season with salt and pepper. Cover with aluminum foil to keep warm.

GRILLED SWEET ONION

1	tablespoon extra virgin olive oil
1	Vidalia or other sweet onion, sliced into ¼-inch rings
1	tablespoon balsamic vinegar

Heat the olive oil in a grill pan or cast-iron skillet over medium-high heat. Add the onions and cook for 5 to 7 minutes; the onions should still have some crunch. Place in a small bowl and drizzle with the balsamic vinegar. Cover the bowl with aluminum foil to keep warm.

LEMON HERB VINAIGRETTE

	Juice of 1 lemon
½	teaspoon honey
½	cup extra virgin olive oil
2	teaspoons fresh thyme, chopped
	Salt and freshly ground black pepper

In a small bowl combine the lemon juice and honey. Whisk in the olive oil in a slow, steady stream until the vinaigrette has emulsified. Add the thyme and salt and pepper to taste.

PAN-SEARED ROCKFISH

4	rockfish (also called striped bass) or red snapper filets, about 8 ounces each
	Salt and freshly ground black pepper
2	tablespoons extra virgin olive oil
2	tablespoons butter
2	sprigs of thyme

Score the skin with a very sharp knife in a crisscross pattern to prevent the fish from curling. Season with salt and pepper.

Heat 2 large skillets over high heat until almost at the smoking point. Add one tablespoon of the olive oil and butter to each skillet along with the thyme sprigs. Place the fish skin-side down and sear each filet approximately 2 to 3 minutes. Flip the filets with a spatula and cook 2 to 3 minutes more. Baste the skin with the butter and thyme sprigs.

To serve, return the grilled onions to one of the pans to heat through, being careful to keep them crisp. Place some of the sweet potato puree in the center of a plate. Top it with the onions and then a filet. Drizzle the vinaigrette on and around the fish. Optional: garnish with tossed micro greens.

AGRICULTURE 101 SOME FARM TERMS DEFINED

Agritourism Agricultural activity that brings visitors to a farm, including farm tours, farm stays, and workshops.

Biodynamic farming A holistic approach to agriculture associated with the Austrian philosopher and social thinker Rudolf Steiner. It relates the ecology of the earth-organism to that of the entire cosmos. All biodynamic farms are by nature organic.

Broilers A chicken fit for broiling, especially a tender young one.

Century farm A program the North Carolina Department of Agriculture started in 1970 to identify families who have owned or operated a farm in the state for 100 years or more.

Conservation easement A restriction placed on a piece of property to protect it from development.

CSA Community Supported Agriculture, where a farmer sells "shares" to consumers who prepay for a season of produce or other goods.

Farmstead cheese Cheese made on a farm, usually in relatively small batches, from milk produced on the farm where the animals are raised.

Grass-fed beef Generally speaking, beef from cattle that have eaten only grass or forage throughout their lives and are not confined.

Heirloom varieties Older types of plants not used in industrial agriculture.

Heritage livestock breeds According to the American Livestock Breeds Conservancy, varieties developed before 1925 and bred continuously since.

Laying hens Chickens used primarily for egg production.

Muscadine A broad category of grape that includes many varieties of bronze and black grapes. Cultivated since the sixteenth century, it is native to the Southeast.

Permaculture A growing method that uses existing relationships within and between plant and animal life to regulate and promote food growth, emphasizing the preservation of the environment.

Row crops Crops grown in a row and typically farmed with machinery, including peanuts, cotton, soybeans, and tobacco.

Scuppernong grape A bronze variety of a muscadine grape.

Sorghum syrup Called molasses by many people but made from sorghum cane. Molasses is a byproduct of crystallizing sugar.

Sustainable A method of harvesting or using a resource in which the resource is not depleted or permanently damaged.

Truffles Edible fungi that grow underground on or near the roots of trees and are considered delicacies.

Vinifera Grape from a cultivated variety of the common grape vine of Europe.

COUNTY-BY-COUNTY LISTINGS

Alamance
Benjamin Vineyards and Winery
Cane Creek Farm
Company Shops Market
Herb Haven
Iseley Farms
Saxapahaw General Store and Café
Saxapahaw Rivermill Farmers' Market
The Winery at Iron Gate Farm

Alexander
Deal Orchards

Alleghany
Crosscreek Farm
Joe Edwards Christmas Tree Farm
Papa Goat's Tree Farm

Anson
Pee Dee Orchards

Ashe
Ashe County Farmers' Market
Big Horse Creek Farm
Lil' Grandfather Mountain Christmas Tree Farm
Mistletoe Meadows

Old Orchard Creek
Shady Rest Tree Farm
Spin a Yarn
Wayland's Nursery
West End Choose and Cut and West End Wreaths
Zydeco Moon Farm and Cabins

Avery
Banner Elk Winery
Elk River Evergreens
Evergreen Ridge Choose and Cut Farm
Franklin Tree Farm
Sugar Plum Farms
Sundance Farm

Beaufort
Southside Farms
Terra Ceia Farms

Bertie
Bertie County Peanuts

Bladen
Lu Mil Vineyard

Brunswick
Holden Brothers Farm Market
Indigo Farms
Lockwood Folly Marketplace
Shelton Herb Farm

Buncombe
The Admiral
Asheville City Market
Biltmore Winery
The Blackbird
Black Mountain Farmers Market
Black Mountain Tailgate Market

Blue Ridge Bison
Cloud 9 Farm
Dogwood Hills Farm
Double "G" Ranch
Early Girl Eatery
Earthaven Ecovillage
Earth Fare
Farside Farms
Flying Cloud Farm
French Broad Chocolate Lounge
French Broad Food Co-op
Gladheart Farms
Greenlife Grocery
The Green Sage
The Grove Park Inn Resort and Spa
Hawk and Ivy Bed and Breakfast
Hickory Nut Gap Farm
Imladris Farm
Laurey's Catering and Gourmet to Go
Log Cabin Cooking and Music
Long Branch Environmental Education Center
Mamacita's
The Market Place
Mountain State Fair
Ox-Ford Farm Bed and Breakfast Inn
Round Mountain Creamery
Sandy Hollar Farms
Table
Trout Lily Market
True Nature Country Fair
Tupelo Honey Café
Venezia Dream Farm
Warren Wilson College Farm
West End Bakery and Café
Zambra

Burke

Apple Hill Orchard and Cider Mill
Millstone Meadows Farm

Cabarrus

Barbee Farms
Creekside Farms
Elma C. Lomax Incubator Farm Park
Maple Lane Homestead
Riverbend Farm
Twelve Acre Academy
Yesterways Farm

Carteret

Garner Farms

Caswell

Baldwin Family Farms
Little Meadows Farm
Sleepy Goat Farm
Yancey House Restaurant

Catawba

BirdBrain Ostrich Ranch
Santa's Forest

Chatham

Angelina's Kitchen
Ayrshire Farm
Bluebird Hill Farm
Central Carolina Community College Land Lab
Chatham Marketplace
Cohen Farm
Fearrington House Restaurant
Jordan Lake Christmas Tree Farm
The Inn at Celebrity Dairy
Piedmont Biofarm

Cherokee
Calaboose Cellars

Chowan
Wilbur R. Bunch's Produce Stand

Clay
Bedford Falls Alpaca Farm
John C. Campbell Folk School

Cleveland
HarvestWorks
Knob Creek Farms

Columbus
Grapefull Sisters Vineyard
North Carolina Strawberry Festival

Craven
A Day at the Farm
Moore Christmas Tree Farm
'R Garden

Cumberland
Gillis Hill Farm
Sandhills Farmers Market
West Produce

Currituck
Rose Produce and Seafood Market
Weeping Radish Farm Brewery

Dare
Island Farm
The Sanderling Resort and Spa

Davidson

Denton FarmPark

SandyCreek Farm

Davie

Ijames Heritage Farm

RayLen Vineyards

Duplin

A. J. Bullard's Orchard

Beautancus Christmas Tree Farm

Duplin Winery

North Carolina Muscadine Harvest Festival

Tarkil Branch Farm's Homestead Museum

Durham

Duke Homestead State Historic Site

Durham Farmers' Market

Elodie Farms

Four Square Restaurant

Fullsteam Brewery

Ganyard Hill Farm

Herndon Hills Farm

LocoPops

Magnolia Grill

Piedmont

Prodigal Farm

The Refectory Café

Scratch Baking

SEEDS

Watts Grocery

Forsyth

Clod-Buster Farms

Dixie Classic Fair

Krankies Local Market

Meridian Restaurant

Noble's Grille
North Carolina Wine Festival
Westbend Vineyards
Winston-Salem Downtown Farmer's Market

Franklin
Franklin County Farm Tour
Hill Ridge Farms
Lynch Creek Farm
Turtle Mist Farm
Vollmer Farm

Gaston
Apple Orchard Farm
Lewis Farm
Lineberger's Maple Springs Farm
Stowe Dairy Farms

Graham
Stoney Hollow Farm
Yellow Branch Farm and Pottery

Granville
Lyon Farms
Triple B Farms

Greene
Benjamin W. Best Inn and Carriage House
Dail Family Produce
Rainbow Meadow Farms

Guilford
Bistro Sofia
Early Farms
The Edible Schoolyard at Greensboro Children's Museum
Greensboro Farmers' Curb Market
Homeland Creamery

Ingram's Strawberry Farm
Lucky 32 Southern Kitchen
Piedmont Triad Farmers Market
Rudd Farm
Steeple Hill Farm
Sticks and Stones
Sweet Basil's

Halifax
Halifax County Harvest Days Festival

Harnett
Coats Farmer's Day
Touchstone Energy North Carolina Cotton Festival

Haywood
Annual Ramp Convention
Boyd Mountain Christmas Tree Farm
The Gardener's House at Frog Holler
Haywood's Historic Farmers' Market
Sunburst Trout Company
The Ten Acre Garden
Waynesville Tailgate Market
Wildcat Ridge Farm

Henderson
Carl Sandburg Home
Edmundson Produce Farm Market
Flat Rock Tailgate Market
Henderson County Curb Market
Henderson County Tailgate Market
Hendersonville Community Co-op
J. H. Stepp Farm's Hillcrest Orchard
Justus Orchard
Lyda Farms
North Carolina Apple Festival
Piney Mountain Orchards Produce
Sky Top Orchard

Square One Bistro
Windy Ridge Organic Farms

Hoke
G. R. Autry and Son Farm and Peach Orchard
John L. Council Farms
Raft Swamp Farms

Iredell
Carrigan Farms
Mills Garden Herb Farm

Jackson
Guadalupe Cafe
Jackson County Farmers' Market
Ty-Lyn Plantation

Johnston
Benson Mule Days
Hinnant Family Vineyards
Lazy O Farm
Smith's Nursery and Produce Farm
Tobacco Farm Life Museum

Jones
Scott Farm Organics

Lee
Gross Farms

Lenoir
Chef & the Farmer

Lincoln
Grateful Growers Farm
Helms Christmas Tree Farm
WoodMill Winery

Macon

Deal Family Farm
J. W. Mitchell Farms
Spring Ridge Creamery

Madison

Bend of Ivy Lodge
Briar Rose Farm
Broadwing Farm Cabins
Doubletree Farm
Eagle Feather Organic Farm
East Fork Farm
Good Stuff
Madison County Farmers and Artisans Market
Madison Family Farms
Spinning Spider Creamery
Wake Robin Farm Breads
Zimmerman's Berry Farm

Martin

Big Mill Bed and Breakfast
Carolina Country Fresh
Mackey's Ferry Peanuts

McDowell

The Cottages at Spring House Farm
The Historic Orchard at Altapass
Meadowbrook Nursery/We-Du Natives
Peaceful Valley Farm
South Creek Vineyards and Winery

Mecklenburg

Artisan
Barrington's Restaurant
The Bradford Store
Charlotte Regional Farmers Market
The Davidson Farmer's Market
Flatiron Kitchen and Tap House

Global
Historic Latta Plantation
Matthews Community Farmers' Market
Renfrow Hardware and General Merchandise
Rooster's Wood-Fired Kitchen
Rural Hill

Mitchell
Harrell Hill Farms
Knife & Fork
Laurel Oaks Farm
OakMoon Farm and Creamery

Montgomery
Johnson Farm

Moore
Ashten's Restaurant
Auman Orchard
Chappell Peaches and Apples
Crystal Pines Alpaca Farm
Doby Christmas Tree Farm
Highlanders Farm
Kalawi Farm
Malcolm Blue Farm and Museum
Moore County Farmers Market

Nash
Bailey's Berry Farm
Finch Blueberry Nursery
Fisher Pumpkin Farm
Nash County Farmers Market
Wrenn Farm

New Hanover
Catch
Farmers' Market at Poplar Grove
Lewis Nursery and Farms

Riverfront Farmers' Market, Wilmington
Tidal Creek Cooperative Food Market

Onslow
Mike's Farm, Country Store, and Restaurant
Onslow Farmers Market
Richlands Farmer's Day
Whispering Dove Goat Ranch and Apiary

Orange
Anathoth Community Garden
Avillion Farm
Captain John S. Pope Farm
Carrboro Farmers' Market
Chapel Hill Creamery
Coon Rock Farm
Crook's Corner
Fickle Creek Farm
Garland Truffles
Lantern
Maple View Agricultural Center
Maple View Farm Country Store
Margaret's Cantina
McAdams Farm
McKee's Cedar Creek Farm Cornfield Maze
Neal's Deli
Niche Gardens
Panciuto
Panzanella
Pickards Mountain Eco-Institute
A Southern Season
Spence's Farm
Sunset Ridge Buffalo Farm
Vista Wood Bison Ranch
Walters Unlimited at Carls-Beth Farm
W. C. Breeze Family Farm Extension and Research Center
Weaver Street Market

Whitted Bowers Farm
Woodcrest Farm

Pasquotank County
Downtown Waterfront Market

Pender
Bannerman Vineyard/White Oak Farm
Nature's Way Farm and Seafood
North Carolina Blueberry Festival

Perquimans
Springfield Bed and Breakfast

Person
Sunset Ridge Buffalo Farm

Pitt
Ayden Collard Festival
Briley's Farm Market
Renston Homestead

Polk
Apple Mill
Beneficial Foods Natural Market
Giardini Trattoria and Giardini Pasta and Catering Company
Green Creek Winery
Green River Vineyard Bed and Breakfast
The Manna Cabanna
The Purple Onion Café and Coffeehouse
Rockhouse Vineyards

Randolph
Caraway Alpacas
Goat Lady Dairy
Millstone Creek Orchards
Rising Meadow Farm

Richmond

The Berry Patch

Triple L Farm

Robeson

Geraldine's Peaches and Produce

Rockingham

Century Farm Orchards

Cornerstone Garlic Farm

Haight Orchards

Hills and Hollows Farm and Museum

Pennwood Farm

Rockingham County Farmers' Market

Tuttle's Berry and Vegetable Farm

Wild Foods Weekend

Rowan

Fisher Farms

Patterson Farm

Rutherford

M^2 Restaurant

Sampson

B&B Pecan Processors of North Carolina/Elizabeth's Pecans

Jackson Farm

Scotland

John Blue House and Cotton Gin

Stanly

Dennis Vineyards Winery

Laughing Owl Farm

Uwharrie Vineyards

Stokes

Horne Creek Living Historical Farm

Keep Your Fork Farm

Surry

Carolina Heritage Vineyards

Elkin Creek Vineyard

Grassy Creek Vineyard and Winery

McRitchie Winery and Ciderworks

Round Peak Vineyards

Shelton Vineyards

Stony Knoll Vineyards

Yadkin Valley Wine Festival

Swain

Cherokee Farmer's Tailgate Market

The Cottage Craftsman

Darnell Farms

Mountain Farm Museum

Transylvania

Queens Produce and Berry Farm

Transylvania County Tailgate Market

Union

Aw Shucks!

The Inn at New Town Farms

Poplar Ridge Farm

Wake

Back Achers Christmas Tree Farm

Got to Be NC Festival

Green Planet Farm

Hen-side the Beltline Tour d'Coop

Herons

Hilltop Farms

Historic Oak View County Park

The Little Herb House

Midtown Farmers' Market
North Carolina State Fair
Poole's Downtown Diner
Pop-n-Son Christmas Trees
State Farmers Market
Tomatopalooza
Wake Forest Farmers' Market
Zely & Ritz

Watauga

Apple Hill Farm
Faith Mountain Farm
The Farm at Mollies Branch
Gamekeeper
Goodnight Family Sustainable Development Teaching
    and Research Farm
J & D Tree Farms
The Mast Farm Inn
The Old Farmhouse
Peak Cut Flower Farm
Songbird Cabin
Swinging Bridge Farm
Watauga County Farmers' Market
What Fir!

Wayne

Governor Charles B. Aycock Birthplace State Historic Site

Wilkes

Apple Brandy Beef
Brushy Mountain Bee Farm
Four Winds Berry Farm
Raffaldini Vineyards
Tumbling Shoals Farm

Wilson

Deans Farm Market

Yadkin

Divine Llama Vineyards
Flint Hill Vineyards
Hanover Park Vineyard
Laurel Gray Vineyards
RagApple Lassie Vineyards
Sanders Ridge Winery and Restaurant

Yancey

EnergyXchange's Project Branch Out
Maple Creek Farm
Mountain Farm
Wellspring Farm

Mountains Multicounty

The Family Farm Tour
High Country Farm Tour

Triad Multicounty

Yadkin Valley Wine Tours

Triangle Multicounty

Piedmont Farm Tour and Eastern Triangle Farm Tour
Taste Carolina and Triangle Food Tour
Whole Foods Market

RESOURCES WHERE TO GO FOR MORE INFORMATION

There are hundreds of publications, films, and websites concerned with farms, farmers, farmland, and fresh food. Here are a few of my favorites.

BOOKS AND MAGAZINES

Barbara Berst Adams, *The New Agritourism* (New World Publishing, 2008)

Sharon Astyk and Aaron Newton, *A Nation of Farmers* (New Society Publishers, 2009)

John T. Edge, ed., *Foodways*, vol. 7 of *The New Encyclopedia of Southern Culture* (University of North Carolina Press, 2007)

Edible Piedmont magazine

Jonathan Safran Foer, *Eating Animals* (Little, Brown and Company, 2009)

David E. Gumpert, *The Raw Milk Revolution* (Chelsea Green Publishing, 2009)

Barbara Kingsolver, *Animal, Vegetable, Miracle* (Harper Collins Publishers, 2007)

Joseph Mills and Danielle Tarmey, *A Guide to North Carolina's Wineries* (John F. Blair, 2007)

Michael Pollan, *The Omnivore's Dilemma* (Penguin Press, 2006)

William S. Powell, ed., *The Encyclopedia of North Carolina* (University of North Carolina Press, 2006)

FILMS

Food, Inc. (directed by Robert Kenner, 2008)
Fresh (directed by Ana Sofia Jones, 2009)
The Garden (directed by Scott Hamilton Kennedy, 2008)
King Corn (directed by Aaron Woolf, 2007)
One Man, One Cow, One Planet: How to Save the World (directed by
    Thomas Burstyn, 2007)
The Real Dirt on Farmer John (directed by Taggart Siegel, 2005)
The World According to Monsanto (directed by Marie-Monique Robin,
    2008)

WEBSITES

Farmworker Rights
www.floc.com
www.ncfarmworkers.org
www.nfwn.org

Hunger, Nutrition
www.endhunger.org
www.farmtoschool.org
www.gardenwriters.org

National Directories
www.eatwellguide.org
www.localharvest.org
www.pickyourown.org
www.rodaleinstitute.org/farm_locator

North Carolina Directories and Information
www.buyappalachian.org
www.ces.ncsu.edu
www.gottobenc.com
www.handmadeinamerica.org
www.homegrownhandmade.com
www.ncapples.com

www.nc-chooseandcut.com

www.ncchristmastrees.com

www.ncfarmfresh.com

www.ncfb.org

www.ncwine.com

www.quilttrailswnc.org

www.slowfoodcharlotte.org, www.slowfoodtriangle.org,
    www.slowfoodpiedmont.org

www.visitnc.com

www.visitncfarms.com

www.visitncwine.com

Promoting Family Farms and Sustainable Farms

www.albc-usa.org

www.animalwelfareapproved.org

www.asapconnections.org

www.attra.org

www.bfaa-us.org

www.carolinafarmstewards.org

www.certifiedhumane.org

www.farmforward.com

www.farmland.org

www.farmtoschool.org

www.growingpower.org

www.growingsmallfarms.org

www.ncchoices.com

www.sare.org

www.slowfoodusa.org

www.southernfoodways.com

www.ssawg.org

www.toxicfreenc.org

www.wwoof.org

GIVING THANKS

Many fine folks helped this book from seed to harvest. First, the farmers. They welcomed me onto their land and into their homes, allowed me to take photos of them (even when they were sweaty and dirty, if I pleaded), shared their stories and sometimes even their harvests. They have my undying gratitude and respect.

The annual Piedmont Farm Tour first raised my awareness of our family farms. Thanks to the Carolina Farm Stewardship Association for starting that way back in 1995, and for providing me with bushels of useful information. Peter Marks and his colleagues in the Appalachian Sustainable Agriculture Project made my western research much easier, especially with their Local Food Guide. Martha Glass, manager of the Agritourism Office at the North Carolina Department of Agriculture, was an invaluable resource, as were the directories she publishes. Too many county and state extension agents to mention passed along information and insight, but I must single out Debbie Roos and Jeanine Davis, who generously share their vast knowledge with the public.

The Golden LEAF Foundation assisted the production of this book in the form of a Local Foods Initiative grant to my publisher, the University of North Carolina Press. I am deeply honored that the foundation felt that Farm Fresh North Carolina would benefit farmers and farming communities economically.

Tourism officials across the state, especially Wit Tuttle, Susan Dosier, and Alex Naar, provided invaluable assistance, as did Brian Long in the Department of Agriculture. A tremendous thanks goes to all the fine establishments that donated my lodging. They are Broadwing Farm, Hot Springs; Ponder Cove, Mars Hill; Jackson Farm, Godwin; Fickle Creek

Farm, Efland; Mast Farm Inn, Valle Crucis; Comfort Inn, Pinehurst; Big Mill Inn, Williamston; Springfield B&B, Hertford; Boyd Tree Farms, Waynesville; Venezia Dream Alpaca Farm, Asheville; Squire's Vintage Inn, Duplin; Town of Surf City; Best Western, Swansboro; Charlotte Regional Visitors Authority; Cabarrus County Convention and Visitors Bureau; Kerr House B&B, Statesville; Hampton Inn, Wilkesboro; Holiday Inn Express, Shelby; Hampton Inn, Marion; Days Inn, Columbus; Hampton Inn, Hendersonville; Microtel, Franklin; Hinton Rural Life Center, Hayesville; and Comfort Suites, Winston-Salem.

To my 1994 Honda Civic hatchback, Garmin GPS, Google maps, and North Carolina DeLorme Gazetteer, without you I couldn't have reached the hundreds of addresses I did in the 23,000 miles I drove for this book.

Support from fellow writers and friends is immeasurable. Head cheerleaders Chuck Adams and Michael Taeckens dispensed much professional advice and kept their pompoms shaking from start to finish. Cheering on the sidelines were Anne Bramley, Sidney Cruze, Kelley Griffin, Carol Kline, Judy Martell, and "Open Hearts, Open Minders." Special thanks to author Nan Chase, who opened her home to me sight unseen. I'm also indebted to food and wine writers across the state, whose work I stumbled onto time and again during my research, especially Joseph Mills, Amber Nimocks, Ann Prospero, Kathleen Purvis, Hanna Rachel Raskin, Danielle Tarmey, Fred Thompson, and Andrea Weigl. I'm grateful to chef Drew Brown in Durham for deepening my understanding of farm-to-fork practices.

A tip of the chef's hat goes to the farmers, cookbook writers, and chefs who donated the recipes in this book. You all have great taste. Every book with recipes needs a recipe tester. I'm lucky to be a friend and neighbor of a longtime pro. Not only did Wendy Goldstein do a remarkable job of turning a jumble of recipes into tasty dishes every home cook can make; she and her family kept me laughing every time I stopped by for a tasting or just to visit.

I took such an immediate liking to Elaine Maisner, my editor at the University of North Carolina Press, that I was worried it would wear off. It didn't. She is whip-smart, wicked funny, and can turn the clunkiest sentence into something elegant and sophisticated. Thanks also to Tema Larter, Paula Wald, UNC Press's exceptional marketing team, and everyone else in the office who had a hand in the birthing of this book. And a deep bow to Alex Martin, copyeditor extraordinaire, who saved me from myself many times over. You are all pros, and your contributions are deeply appreciated.

Keeping things in the UNC family, I turned to the university's School of Journalism and Mass Communication when looking for fact-checking interns. Tremendous gratitude goes to Anna Claire Eddington and especially Sonya Chudgar and Andrea Ludtke, all amazing, ambitious young women who caught more flubs than I'd like to admit.

To my dear friend Kristin Thalheimer, who served as sweep, I send this message: "Love your proofreading generosity—Didy."

The biggest bear hug possible goes to my partner, Lina Kok. She either took or processed every photo in this book or on our website, as well as thousands of others that are meticulously organized in her computer. She endured my many days away from home, tended to the *tekkels*, and kept me going with her encouragement, affection, technical support, and deadline spreadsheets. *Liefde* is love.

INDEX

ABOUT THE AUTHOR

 SELINA KOK

Diane Daniel's first childhood outings were farm tours of a sort, as she traveled the North Carolina countryside with her mother, a public health nurse, on trips to patients' homes. Since then, she has climbed mountains in Ecuador, Argentina, and Indonesia; bicycled Montana's Lewis and Clark Trail; snowshoed backcountry Maine; and kayaked with manatees in Florida. Closer to home, she has camped along the Roanoke River Paddle Trail, paddled around Bald Head Island, cycled much of the Blue Ridge Parkway, enjoyed the fruits of Yadkin Valley wineries, and toured scores of farms in the Carolinas and Virginia.

Since 2002, Diane has been a freelance journalist, writing about her favorite things: the outdoors, travel, and fascinating folks. She's most interested in what she calls preservation travel—travel that preserves cultures, community, history, buildings, and the environment. In addition to travel, Diane's stories have examined community-supported fisheries, food waste, organic Christmas trees, electric cars, and noise pollution. Her work has appeared in the *New York Times*, the *Boston Globe*, the *Washington Post*, *Southern Living*, *Gourmet*, *National Geographic Traveler*, the *Raleigh News & Observer*, *Our State*, *Ode Magazine*, and other papers and magazines. In 2008, she won a national Lowell Thomas Award for her *Times* travel story on regional home exchanges.

Diane lives in Durham, N.C., only a mile from the site where her Granny Fanny served Sunday dinners of fried chicken, string beans, and biscuits. You can learn more about her at www.farmfreshnorthcarolina.com.

Other **Southern Gateways Guides** you might enjoy

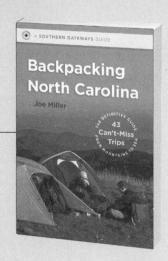

Backpacking North Carolina The Definitive
Guide to 43 Can't-Miss Trips from Mountains to Sea

JOE MILLER

*From classic mountain trails to little-known gems
of the Piedmont and coastal regions*

**A Field Guide to Wildflowers
of the Sandhills Region** North Carolina,
South Carolina, and Georgia

BRUCE A. SORRIE

*The first-ever field guide to the wildflowers of
this vibrant, biodiverse region*

**Wildflowers & Plant Communities of
the Southern Appalachian Mountains
and Piedmont** A Naturalist's Guide to the
Carolinas, Virginia, Tennessee, and Georgia

TIMOTHY P. SPIRA

A habitat approach to identifying plants and interpreting nature